Electronics Explained

Electronics Explained

The New Systems Approach to Learning Electronics

Louis E. Frenzel, Jr.

ELSEVIER

AMSTERDAM • BOSTON • HEIDELBERG • LONDON
NEW YORK • OXFORD • PARIS • SAN DIEGO
SAN FRANCISCO • SINGAPORE • SYDNEY • TOKYO

Newnes is an imprint of Elsevier

Newnes

Newnes is an imprint of Elsevier
30 Corporate Drive, Suite 400, Burlington, MA 01803, USA
The Boulevard, Langford Lane, Kidlington, Oxford OX5 1GB, UK

Library of Congress Cataloging-in-Publication Data
Frenzel, Louis E.
   Electronics explained : the new systems approach to learning electronics / Louis E. Frenzel, Jr.
      p. cm.
   ISBN 978-1-85617-700-9
   1. Electronics—Textbooks.   I. Title.
   TK7816.F683 2010
   621.381—dc22   2010006565

British Library Cataloguing-in-Publication Data
A catalogue record for this book is available from the British Library.

For information on all Newnes publications
visit our web site at www.elsevierdirect.com

Typeset by MPS Limited, a Macmillan Company, Chennai,
India www.macmillansolutions.com

Printed in the United States of America

10 11 12 13 14   10 9 8 7 6 5 4 3 2 1

Working together to grow
libraries in developing countries

www.elsevier.com | www.bookaid.org | www.sabre.org

ELSEVIER   BOOK AID International   Sabre Foundation

With love to my one and only, Joan Ree.

Contents

3. The Systems versus Components View of Electronics

4. Electronic Circuits: Linear/Analog

5. Electronic Circuits: Digital

6. How Microcomputers Work

7. Radio/Wireless

10. Audio Electronics

11. Video Technology

List of Figures

Preface

This book is for those of you who want to learn about electronics and computers but frankly are put off by all the gory details. If you have a short attention span, a nonexistent background in science, math, and technology, and are looking for near-instant gratification of electronics knowledge, this book is for you. Electronics is a complex subject, especially if you want to be an electronics engineer. However, if you are merely curious or just need a working knowledge of electronics for your job, this book will give you a basic education on the fundamentals without boring you to death or loading you down with the excruciating details that other books lay on you. What I have attempted to do is write a book that tells you the most important concepts and fundamentals but leaves out the minutiae that only an engineer will need or appreciate. This book contains exactly what you need to know today in the 21st century.

There are lots of basic electronics books around and you may already have read some. This one is really different. First, it does not go into detailed circuit analysis. Why? Simply, all electronic equipment today is made up of integrated circuits or chips, and we can't see the circuits, access them, or even repair them. If you are an electronic engineer designing integrated circuits, yes, you do need to understand their operation. If you just want to learn how electronic equipment works, then you can skip this detail. This book treats circuits as simple functional building blocks whose purpose and application are easily understood.

Second, this book presents what I call the NEW electronics. Most basic electronic books dwell on the older components and circuits and ignore how electronics really is today. These books do a good job teaching you the history of electronics but little else. What you will see in this book is a fresh new way to look at and understand electronics, especially in the context of the familiar applications you use every day made with the most advanced integrated circuits. If you complete this book and still want more, one of the more traditional basic electronics books would be your next step. See the appendix for my recommendations.

The first three chapters introduce you to the electrical principles that form the basis of electronics. You will learn the core concept of electronics, which boils down to how to generate current flow and then control it to do something useful. Magnetic fields are also covered, since magnetism is just as important in electronics as current flow. But don't worry, even if you never took a physics course you can easily get through this material.

Next, you learn about the basic components of electronics such as resistors, capacitors, inductors, transformers, diodes, transistors, and especially integrated circuits. Then you will discover all the different types of electronic circuits used to make up any electronic device. The functional block diagram approach is used here so you won't have to do the nitty-gritty circuit analysis that has you chasing electrons through components and solving simultaneous equations.

An important part of this book is the coverage of computers, and specifically of embedded controllers, that is, those tiny single-chip microcomputers or microcontrollers that are built into virtually every electronic device. Digital signal processors (DSPs) are also covered since they also appear in most modern electronic equipment. You can't name an electronic product today that does not have one of these embedded computers in it. You really need to know what these are and how they work, as they are the very heart of electronic devices today.

The rest of the book consists of descriptions of how all the major electronic products and systems work. The entire range of modern electronics is covered, including major things like cell phones, TV (cable, HDTV, DVD, VCR, etc.), audio (stereo, CDs, MP3, etc.), satellites (GPS, TV, radio, etc.), radio (broadcast, marine-aircraft-shortwave, ham, CB, family radio, etc.), and everything else you can think of. And it is all explained using basic building blocks rather than detailed circuits.

Finally, if you want involvement as you read the book, get involved with the projects that I list in the Project sidebars in each chapter. These can lead to a deeper understanding of the concepts. Furthermore, these projects could encourage you to become a true electronic hobbyist ... or is that a geek or nerd?

When you finish this book you will not be an electronic engineer nor will you be able to design or analyze circuits. But you will know about all the different kinds of circuits, how they work, and how they all fit together to create electronic equipment. The benefits to you are the ability to apply, use, select, operate, and to some extent troubleshoot most of the common electronics products with confidence, while appreciating their complexity and the value they deliver. Hope you like it.

Lou Frenzel
Austin, TX

Chapter 1

Learning about Electronics for Your Own Good

A Little Bit of Perspective before You Get to the Details

In this Chapter:
- Why it is helpful to know electronics.
- The 10,000-foot view of the electronics industry and how it works.
- The systems view of electronics.

INTRODUCTION

You don't know how lucky you are. You live in high-tech heaven surrounded by lots of useful, entertaining, interesting, necessary, and even addicting electronic products, products you cling to and would not want to give up.

But do you have any idea how many electronic products you own? Do you really know how much electronics influences your life every day? Some of you do, but most of you probably do not because your electronic gadgets are pretty much taken for granted. Yet we spend the better part of our lives working with these devices. Most of us sit in front of a computer all day at work, slouch on the couch in front of the TV at night, listen to our iPods going to and from work, and spend a weekend downloading videos off of the Internet or taking photos with a digital camera. And let's be clear here, how many text messages did you send today or cell calls did you make? The impact on our lives of all this electronic equipment is almost overwhelming. In this first chapter, I encourage you to think about how electronics impacts your life.

And that brings us to the subject of this book, learning how electronics works. If you want a first book in electronics that gives you the big picture about how electronic equipment works, this is it. I am not going to beat you to death with a lot of complex circuits and theory or math, only what you really need to know. Then I will go on to show you how each and every one of the products you use every day actually works.

doi: 10.1016/B978-1-85617-700-9.00001-1

IT ISN'T LIKE IT USED TO BE

It used to be that you learned electronics by playing around with resistors, capacitors, transistors, and other devices. You wired up simple circuits, built a one- or two-transistor radio, or even built a kit. But today electronics is so different that it is hard to do that anymore. Oh yes, you can still build little circuits and radios and other gadgets, but that is not how electronic products are made today. Electronic equipment today is made with integrated circuits or ICs that we also call chips. These chips are tightly packed on printed circuit boards with resistors and capacitors that you can barely see. It is more difficult than ever to experiment with these devices, much less troubleshoot and repair them. We are in that age of electronics where unless it is a big pricey piece of equipment, you throw a defective electronic product away and simply buy a newer and better one at a lower price.

But despite the complexity of today's electronic products, there is something just fascinating about them. Many of you actually want to know how they work. And despite their complexity, these products are actually understandable. It is possible to learn how these devices work without a whole lot of trouble. And you don't have to learn all the math and physics required by engineers to do so. This book takes the basic ideas of electronics and reduces them down to the lowest common denominator and shows you how to use those fundamentals and apply them to all of the fabulous electronic gadgets you have today.

WHAT'S IN IT FOR YOU?

There are some big benefits to being electronic literate. Take a look at some of them.

Satisfy Your Curiosity—Surrounded by so much electronics, you often probably wondered just how all this stuff works. For some of us, not knowing how something works nags at us until we satisfy that curiosity. If you have an interest in electronics, you tend to want to know more about it. This book will get you started in the right direction.

Enhance Your Hobby—If you dabble in a hobby that uses electronics, you obviously want and need to know more about it. Lots of hobbies involve electronics such as radio-controlled models, electric trains, amateur radio, shortwave listening, citizens or family radio, audio systems, or video. Or maybe you just like playing around with your home security system, garage door opener, or your high-speed Internet connection. A major part of any hobby is learning more about it and that is very satisfying. And don't forget—electronics itself can be your hobby. You can spend your time learning more about it and then digging deeper into those things that you want to know more about. Amateur radio is certainly one of those electronic hobbies that will take you from one extreme to the next in the radio spectrum. Anything having to do with computers is also a great hobby.

With everything being controlled by an embedded microcomputer, spending more time learning about them is a great activity with nearly endless possibilities.

Job-Related Needs—You may actually be working in a job surrounded by electronic equipment of some sort, yet you do not know how it works. Often, just understanding some of the fundamentals will give you a better grasp of how to use that equipment or even troubleshoot and repair it. Understanding some of the concepts may actually help you to better select and purchase equipment that you may use.

Boost Your School Performance—For those of you going to a college or university to learn electronic engineering or technology, you regularly have your curiosity stimulated by the classes you take. Yet most of the classes are heavily involved with theory, physics, math, and circuit analysis. What you really don't get in school is the perspective or big picture as to how the electronic products actually work. This book will give you the big picture and show you how it all comes together so that all those detailed classes in circuits and theory will make more sense.

Just so You Won't Appear to Be So Dumb—With electronics everywhere, people often talk about it without really knowing what they are talking about. It is amazing the ignorance of people who use the electronics but don't have a clue as to what it does or how it does it. For instance, do you really understand what an MP3 is, or why BluRay DVDs are better, or what a 3G cell phone really is? The answers are in this book. If you want to appear a little smarter to your co-workers, friends, and relatives, this book can help out. You will be able to deal with common questions and misconceptions, such as can you get shocked by a car battery, can you really get brain cancer or pop popcorn with your cell phone, or will sitting too close to your TV set make you sterile? While this book certainly won't make you a genius, it will indeed give you the basics so that you won't appear to be an idiot when technical questions about electronics come up.

ELECTRONICS: THE BIG PICTURE

Before we get into the nitty-gritty of how electronic equipment works, you really need to know the entire scope of electronics. It is a huge industry and recent figures seem to show that it is one of the biggest, if not the biggest, markets in the entire world. To get a grasp of how big the industry is, we normally break it down into segments of specializations. There are five major fields of electronics: communications, computers, control, instrumentation and measurement, and components. Any electronic product or device will easily fall into one of those categories. In some cases, the product may overlap two or more or all of those categories. Take a look at Figure 1.1, which shows each of these segments of electronics and the products and technologies that are part of them.

Communications	Computer	Control	Instrumentation	Components
Radio	Supercomputers	Automation	Test equipment	Resistors
Television	Mainframes	Robots	Test systems	Capacitors
Telephone	Servers	Manufacturing	Data acquisition	Inductors
Cable TV	Workstations	Automotive	Medical	Transformers
Satellites	Personal computers	Home control	Auto diagnosis	Diodes
Networks	Peripheral equipment	Appliances	Aircraft diagnosis	Transistors
Wireless systems	Embedded	Security		Integrated circuits
Consumer	Special purpose	Toys		Printed circuits
Cell phones				Wire and cable
				Connectors

FIGURE 1.1   Major sectors of the electronics industry and common applications.

Communications

Communications is the oldest segment of electronics and still the largest. Electronics actually started with communications–specifically telegraph, telephone, and radio–and grew from there. A lot of this book is about communications because it is such a huge segment of the industry and one that we use the most. And don't forget, communications comes in both wired and wireless forms. Wired systems include the vast telephone network, cable TV systems, and computer networks, including the Internet. Wireless systems comprise all of the radio, cell phone, television, satellite, radar, and other wireless systems.

Computers

Computers have only been around since the late 1940s and early 1950s but what an impact they have had on the way we work. Today virtually everyone uses and even owns a personal computer. And then there are all the larger and more powerful computers such as the servers that manage all of our networks, the mainframes still used by government and big business to manage huge databases, as well as the high-speed supercomputers that still do an amazing amount of scientific, engineering, and mathematical design and research. But the computer we are going to talk about most in this book is the embedded controller. These are small single-chip digital computers called *microcomputers* or just *micros* that are literally part of every electronic product. These are miniature digital computers dedicated to a specific function inside the products in which they exist.

Control

Control is a broad general term for monitoring and control. Monitoring, of course, means sensing various physical characteristics such as temperature, humidity, physical position, motor shaft speed, or light level. Electronic devices called *sensors* or *transducers* convert these physical characteristics into electrical signals which allow us to monitor them. Control refers to managing and exercising some degree of influence over items such as motors, lights, relays, heating elements, and other devices. Most monitoring and control functions take place in factories, chemical plants, refineries, and food processing operations. Control also occurs in the home, an example being your electronic thermostat. It also shows up in electromechanical products like a DVD player. Electronic controls appear in appliances, home control systems, security systems, automobiles, and even toys. Robots are another major segment of the control industry.

Instrumentation

Instrumentation refers to the field involved with testing and measuring electronic equipment and other mechanical or electronic items. It refers to the test instruments like volt meters, oscilloscopes, and spectrum analyzers as well as the large automated test systems used for mass testing and measuring of other devices. Instrumentation refers to data acquisition systems, medical tests and measurements, and a wide range of other products.

Components

All of the electronic products that make up each of the four major segments of the electronic industry are made up of various electronic components. In the past, electronic products were made of the individual discrete components such as resistors, capacitors, diodes, and transistors. Today most electronic equipment is made with one or more integrated circuits surrounded by a sprinkling of those other so-called *discrete* components. No longer can we access most individual components or circuits since they are sealed within integrated circuits or chips. Just keep in mind that there is a whole industry centered on making these electronic components that are in turn made available to the manufacturers of the end electronic equipment in each of the four major categories just described.

HOW THE ELECTRONICS INDUSTRY WORKS

As part of the big picture of learning electronics, you really need to understand how the industry itself works. This is summed up by the simplified block diagram in Figure 1.2. It all starts with the raw materials that the component manufacturers use to create the individual components. For example, sand is refined into silicon, which, in turn, is converted into wafers that are processed into the integrated circuits we use.

The electronic equipment manufacturers design the electronic equipment for the marketplace. It may be a TV set, a cell phone, or a military radar. These companies buy the electronic components and put them together to create the

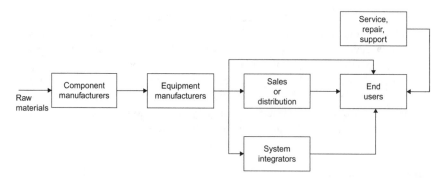

FIGURE 1.2   General block diagram of how the electronics industry works from raw materials to end users.

end product. Manufacturers may then sell the end product directly to customers and end users. In some cases, the end product will go through a sales organization or distributor of some sort. Your LCD large-screen TV set doesn't come directly to you directly from Samsung in Korea or Sony in Japan. It is sold to an organization such as Best Buy or Circuit City, which is where you buy it. Sometimes the electronic equipment is sold to so-called systems integrators. These are companies that piece together larger systems for an end user (like the military) made up of a wide range of different types of equipment.

Finally, there are the customers or end users of the equipment. These come from a wide range of backgrounds. The end users or customers may be businesses with offices, industry with factories and process control plants, or the government, military, hospitals and other medical organizations, schools, colleges and universities, and you the consumer. There is also a network of service companies designed to support customers. They may do installation or repair work that the customer is not capable of doing.

Electricity versus Electronics

One of the questions that always comes up in the early stages of learning about electronics is what the difference is between electricity and electronics. Let's get this straight up front. Electricity refers to the basic energy we use for power. It generally involves the generation and distribution of electrical energy to homes, businesses, and factories. Electricity is energy that is produced by the coal-fired or nuclear power plant and then sent to you over long transmission lines until it comes into your house where you access it through your AC wall outlets. You use that energy for lighting and appliances, and also to power most of your electronic equipment.

Electrical power may also be generated for use in other systems. The electrical system in a car is a great example. It uses a battery plus a generator driven by the automobile engine to recharge the battery and to operate all of the electrical and electronic devices in a car. Ships, airplanes, and off-shore drilling rigs also have their own electrical-power generating systems. The new electric wind generators and solar power systems are also electricity-producing systems.

All of the principles of electricity also apply to electronics. Basic electrical components such as resistors, capacitors, inductors, transformers, motors, and other similar devices are also used in electronics.

Electronics is the field of science that uses electrical principles to perform other useful functions. Electronics is the control of electricity to produce radio, television, computers, robots, and MRI machines. To control the electricity, you need more than just basic electrical components such as resistors and capacitors. What makes electronics possible are the unique components called *transistors*. Electronics began, of course, with vacuum tubes and then switched over to transistors as the main control devices. Transistors are now miniaturized and packaged in devices called *integrated circuits* that perform very specific electronic functions, such as generating signals, amplifying signals, or performing some control function. Just keep in mind that it is the control devices such as transistors and the circuits that they comprise that make electronics what it is. The transistors are used to form circuits that in turn perform all of the various functions required to produce electronic devices.

THE PAINFUL TRUTH

Okay, here comes the hard part. Electricity is moving electrons. An electron is a tiny particle that is inside the atoms that make up everything we can feel, see, and even not see. If we can get some electrons moving together in a wire, electrical or electronic component, or circuit, we can control them to do many useful things, such as amplification, for example. Moving electrons make up what we call *current flow*. And current flow is produced by electrical energy called *voltage*. Voltage comes from batteries, the AC wall outlet, and lots of other sources including power supplies, solar cells, microphones, and the like. If we can control the electrons we can produce magic results. That is what electronics is all about.

We control the electrons with components like resistors, capacitors, inductors, transistors, and all sorts of integrated circuits. We create circuits that actually process the electron flow. We can call the voltages that create current flow and the resulting current *signals*. The circuit processes the signals to produce the end effect that we want, be it a radio, a computer, or DVD player.

One of the basic models I will use in this book is shown in Figure 1.3. Think of electronic signals as the inputs to a circuit or piece of equipment that processes the signals producing new signals that represent the output we want. Think of the small voltage that comes from a microphone when you talk into it. It is basically too small to be useful. So we apply it to a circuit called an *amplifier* that makes the signal bigger and more useful. That amplifier processes the signal to make it large enough to drive a speaker as our desired

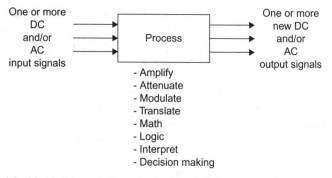

FIGURE 1.3   Model of how all electronic circuits and equipment work. Input signals are processed by circuits and equipment into new output signals.

output. The result is a public address system. Anyway, you get the picture. Just remember that *all* electronic circuits and equipment are based on this model: INPUT-PROCESS-OUTPUT. Think about it.

Here are a couple of other examples to show how most electronic things work.

Cell Phone

Take a look at Figure 1.4. On the left is a cell phone handset. It is made up of two major pieces–a transmitter (TX) and a receiver (RX). You talk into a microphone and generate a small electrical voltage or signal representing your voice. The transmitter takes that and processes it in such a way that the voice controls or modulates a high-frequency radio signal. That signal carrying your voice is then radiated by the antenna to a cell site or base station. The base station receiver picks it up and processes it back into the original voice signal where it then undergoes other processing before being sent over the telephone network to its destination. The dialing process or text messaging from the keyboard also generates input signals that are transmitted to the base station.

On the return side, the voice from your caller passes through the cell phone network to the base station and it is then used to control the transmitter to send that voice to you by radio. The transmitter sends the signal to the antenna that sends the signal to the receiver in your cell phone. The receiver processes the signal by amplifying, demodulating, and otherwise processing the voice that eventually is sent to the speaker that is glued to your ear. Another output may be a text message on the liquid crystal display (LCD).

Computers

The main function of a computer is to store and retrieve data, process data, and to provide automatic control. Data consists of information, facts, and figures. It may be numbers, letters, symbols, or text of any kind. It might represent

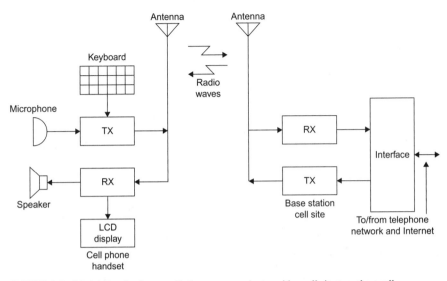

FIGURE 1.4 Model showing how a cell phone communicates with a cell site to make a call.

names and addresses, parts inventory, accounting or banking figures, control signals, or a variety of other items. Data may also be graphics and illustrations, or voice and video that are converted to data. Such data can be stored electronically in computers and rapidly accessed.

Computers process or manipulate data in many different ways. They can perform mathematical operations on it, sort it, edit it, translate it, or otherwise operate upon it to produce some useful output. Computers also make decisions based on the data.

Computers have so many applications that it is difficult to enumerate them. In general, computers are tools that help people organize and perform their work in a more efficient manner. Computers also allow people to do work that was difficult or impossible to do without computers. Computers perform accounting and banking operations; store and process mailing lists and employee records; solve complex mathematical problems in business, science, and engineering; and create graphics. The thing that makes them so valuable is that all of their operations are automatic. Humans need not be a part of the actual computation, as is necessary with a calculator. Computers carry out preprogrammed assignments to solve various problems or perform operations. Most operations are extremely fast. It is this automatic feature that makes computers so useful.

All computers, regardless of their size and features, are made up of the four basic sections shown in Figure 1.5. The memory electronically stores the data to be processed, as well as the program that tells how the data is to be manipulated. The program is a sequence of codes or instructions that cause the computer to carry out its preprogrammed assignment one step at a time.

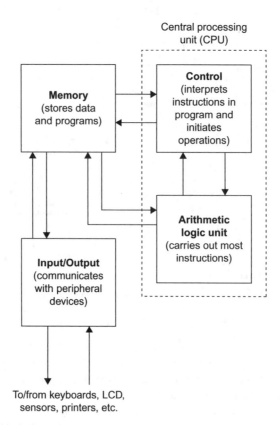

FIGURE 1.5   Block diagram representing any digital computer, be it a mainframe, PC, or embedded controller.

Refer again to Figure 1.5. The control section examines the program instructions one at a time in sequence, interprets them, and then sends control signals to all other parts of the computer. The desired operations, such as store, add, print, and so on, are carried out.

The arithmetic logic unit (ALU) is the section that actually carries out most of the processing. It performs data transfers, arithmetic operations, and logical decision making. In most computers, the control and ALU sections are integral units that work closely together. They are usually referred to as the central processing unit (CPU). A microprocessor in a personal computer is a single chip or central processing unit.

The input/output (I/O) section in Figure 1.5 is the interface between the computer and the outside world. It communicates with peripheral devices that humans use to input programs and data, and receive the output results of computation. Data and instructions are usually entered via a keyboard, while outputs are displayed on a video screen or printed out. Inputs may come from sensors and output may go to a printer or LCD readout.

There are many sizes and types of computers. Supercomputers are the largest and fastest computers. They are used primarily for difficult, complex scientific and math problems. Mainframe computers are also very large, and they are used by business and government for financial applications, data storage, and other applications. Personal computers are the smallest practical computers in use. Millions of them are used for word processing, financial spreadsheets, database management, and email and Internet access.

There are also many dedicated computers, built into other equipment to perform single-purpose operations unlike the general-purpose data processors described above. Thanks to semiconductor technology, it is now possible to put an entire computer on a single chip you can hold in your hand. These are called *embedded computers*, or *microcontrollers*. These microcomputers are used like components and are built into most other kinds of equipment such as TV sets, microwave ovens, gasoline pumps, washing machines, and many others. Virtually every other electronic product contains at least one or more embedded controllers. And even very large ICs have controllers on the same chip called *cores*.

Robots

A great example of a control device is a robot. Most robots consist of an electromechanical manipulator arm used in manufacturing operations. Robots are machines designed to mimic human capabilities. Most robots simulate a human arm and hand. As a result, a robot may be able to substitute for a human in certain manufacturing jobs. A typical robot manipulator arm is shown in Figure 1.6. Motors operate shoulder, elbow, and wrist joints, while grippers and other tools form the hand that does the work.

Robots help automate the manufacturing process to increase productivity. They lower costs, produce more goods in less time, and improve quality. Robots also replace humans in many jobs requiring repetitive, tedious, and hazardous work.

Robots perform a variety of tasks in industry. The most common are welding, spraying or painting, and "pick-and-place" operations. Various tools on the end of the manipulator arm, called *end effectors*, let the robot perform a remarkable variety of tasks. In addition, a programmable controller or computer is used to control the robot. A robot may thus be programmed to perform a wide variety of jobs and is easily reconfigured when the work changes.

In subsequent chapters you will see more of these kinds of descriptions. But in Chapters 2, 3, and 4, you will learn a few more basics that make the rest of the book more understandable.

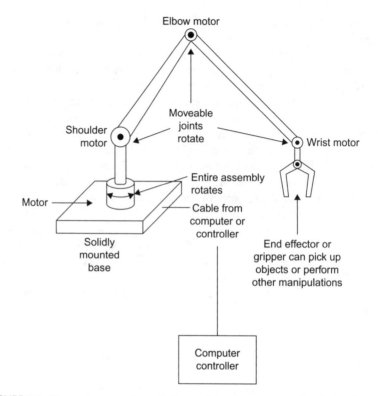

FIGURE 1.6   The most common form of robot emulates a human arm and hand to perform work automatically under the control of a computer.

Project 1.1

How Much Electronics Do You Own?

Just to be sure that you really appreciate the extent to which electronics impacts your life, I strongly suggest that you take a few minutes and make an inventory of all the electronic devices and equipment that you use at home, at work, and away from home. Get a yellow pad and a ballpoint pen and move around your house and just write down all the electronic products that you own and use on a regular basis. TV sets, computers, and cell phones are the most obvious, but don't forget things like telephones, your wrist watch, appliances, and other entertainment products. Make a separate listing of the electrical and electronic products you use inside your car and then another one for those electronic products that you use at work. Make the list comprehensive. Study it. Be amazed.

Project 1.2

How Much Electronics Do You Use?

Now take that same pad of paper and pen and make a quick summary of how you spend your time during the day. How much time do you spend using each of the electronic products that you listed above? Some of them are only momentarily used, but others you use over longer periods of time, particularly computers and TV sets. Just take a quick estimate of the time you use each one during a typical day. Again, add up the hours. Be amazed.

Project 1.3

Decisions, Decisions

Make the following decision. If you had to give up all of your electronic products but one, which one would you keep? TV, cell phone, remote control, iPod? Only one. This will say a lot about who you are.

Chapter 2

Electronic Concepts: More Interesting Than You Think

Some Basic Stuff You Really Need to Know

> **In this Chapter:**
> - Electrons and current flow.
> - DC and AC voltage.
> - Voltage sources, batteries, the power line, and power supplies.
> - Signal sources.
> - Magnetism and induction.

INTRODUCTION

You don't have to be an engineer, physicist, or mathematical genius to learn electronics. But there are a few basic things that you need to know so you can understand what is going on in the circuits and equipment. For example, you need to know that electrons are what makes up current flow, and voltage is what makes the current flow. You also need to know about magnetic fields and electric fields, where they come from, and what they do. And how magnetism and current flow are related. With that background you are ready to learn about the various components and circuits. This chapter takes care of those basics.

ELECTRICITY AND ELECTRONICS

Electricity is a type of energy produced by a charge that is either fixed (static) or moving (dynamic). The main source of electricity is the electron, a subatomic particle that has a negative charge. When electrons are stored or moved, electricity is produced.

Electricity is one of our prime sources of energy. It is used for lighting, heating, and operating appliances of all sorts. It powers motors to produce mechanical energy, and it powers our electronic equipment. The uses of electricity are virtually infinite.

doi: 10.1016/B978-1-85617-700-9.00002-3

15

Electronics is the field of applied science that uses components such as resistors, capacitors, diodes, transistors, and integrated circuits to control and process electricity. These components are used to create circuits that convert, modify, vary, translate, or otherwise manipulate electrical charges to perform useful functions. That is the essence of it. Now for the details.

Atoms and Electrons

This section really ought to be called Physics 101 but I don't want to scare you away. Just to make you more comfortable, I have to say that it is not all that complicated or difficult to understand. So even if you avoided physics in high school, just be aware that it is pretty easy. Trust me.

All matter, whether it is a solid, a liquid, or gas, is made up of tiny entities called atoms. An atom is the smallest possible particle of substances called elements. An element is a chemical substance that cannot be subdivided into smaller, simpler substances. Typical elements are hydrogen, oxygen, gold, silver, copper, carbon, helium, silicon, and sodium. All matter is either an element or composed of several elements. Atoms of two or more elements often combine to form new substances called compounds. For example, water is a compound of hydrogen (H) and oxygen (O) or H_2O. Salt is a compound of sodium (Na) and chlorine (Cl) or NaCl. The smallest possible particle of a compound that exhibits all the characteristics of the substance is called a molecule.

According to the theories of physics, an atom is composed of a nucleus consisting of a core of tightly bound subatomic particles, called protons and neutrons. Protons have a positive charge. Neutrons are neutral, having neither a positive nor a negative charge.

Rotating around the nucleus are electrons. The electrons arrange themselves in multiple orbits in much the same way that the planets orbit around the sun. The orbits are sometimes referred to as rings or shells. Electrons have a negative charge. Figure 2.1 illustrates a copper atom. The number of electrons in orbit equals the number of protons in the nucleus. The number of positive charges equals the number of negative charges. Therefore, the atom is balanced electrically.

As far as electronics is concerned, the most important part of an atom is its electrons. Since electrons can be manipulated (i.e., stored or moved), they can be used to produce electricity. This electricity can then, in turn, be processed and controlled to perform a wide range of valuable functions. By applying an external force, the electrons in the outer shells can be stripped off to produce current flow.

Charge, Voltage, and Current

There are two basic types of electricity, static and dynamic. Static electricity is the buildup of a charge between two objects. Dynamic electricity supplies a continuous charge.

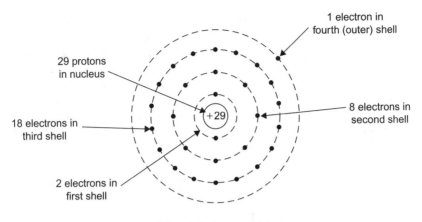

29 protons
in nucleus

1 electron in
fourth (outer) shell

18 electrons in
third shell

8 electrons in
second shell

+29

2 electrons in
first shell

Notes: 1. Protons not shown
2. Number of protons equals number of electrons

FIGURE 2.1   Physics model of what a copper atom looks like.

Static Electricity

With static electricity, one object has an excess of electrons and the other has a shortage of electrons. The object with an excess of electrons is therefore negative while the object with a shortage of electrons is positive. A basic law of electricity states that opposite charges attract and like charges repel. The two oppositely charged bodies, one negative and one positive, will have a high physical attraction for one another. An invisible force field called an *electric field* will exist between the two charged objects. The object with a shortage of electrons attracts the object with excess electrons.

The two oppositely charged objects will naturally attract one another. The closer they are, the greater the attraction. This attraction of the charges sets up an electrical force field between the two bodies. This electric field, like a magnetic field, contains energy that can be released to perform useful work.

Whenever there are opposite charges on two objects, we say that an electrical potential or difference of potential exists between them. This difference of potential is called *voltage*. Voltage is an electrical pressure that can cause work to be accomplished.

The charge stored on the two objects is usually called static electricity because it is stationary. However, this charge is usually dissipated in some way. If the two charged objects get too close, or the charge builds up too high, the attractive force becomes so great that the electrons jump the gap between them and create a spark. That is how lightning occurs. The clouds and the earth become oppositely charged, the earth positive, the clouds negative. The charge becomes so great that the electrons leap across the free space gap, ionizing the air and creating a path through which the electrons travel to neutralize the positive charge on the earth. The result is the massive spark or flash that we call lightning and the roar we call thunder.

In general, static electricity is not very useful. It represents energy stored as a charge, but it quickly dissipates as a spark or a brief flow of electrons between the negatively and positively charged objects when the two charged objects get close enough or when the charge gets too high. We call this an *electrostatic discharge* (ESD). Static electricity, or ESD, is regarded as more of a disadvantage or nuisance than anything else. In fact, it is one of the main causes of damage to electronic components such as transistors and integrated circuits.

Neutralizing a charge between two objects cancels or eliminates the potential. However, if we had a constant charge that could constantly be replenished, we would have a continual source of potential (voltage) and could therefore sustain a flow of electrons. With such a source of electrical energy, useful work can be accomplished.

Dynamic Electricity and Current Flow

Electricity as we know it, whether in electrical power applications or in electronics, is actually the flow of electrons. Electrons flowing from one place to another is known as current flow. And it is voltage that causes the current to flow. When electrons flow, we have energy that can be put to practical use. Electrons flowing in a filament create light. Electrons flowing in a heating coil produce heat. Electrons flowing in a motor produce mechanical energy. The object is to convert the energy from one form to another. In electronics, we precisely control the electrons with special components and circuits to produce a variety of effects.

Figure 2.2 provides a general block diagram model that shows how all electrical or electronic circuits work. A voltage source produces the energy that causes electrons to flow. This current flow is controlled or processed by a single electronic component, or many components that make up an electronic circuit. The processed electrons flow through a load that consumes or uses the electrical energy to produce a useful output. In an electrical application, the load may be a light, a heating element, or motor, as described above. In an electronic application, the load may be an antenna that radiates the radio signal produced by the electronic circuits. Or the load may be the magnetic head that develops and records digital data on a hard disk or a laser that records a video signal on a DVD.

A couple of key facts about the circuit in Figure 2.2 are required here. First, there must be a voltage source to cause the current to flow. The electrons flow from the negative terminal of the voltage source around the circuit as they are attracted by the positive terminal. Second, the circuit must always form a complete loop from plus to minus.

Current is so important that we need some way to measure it. We do this by standardizing the number of electrons that move past one point in a conductor

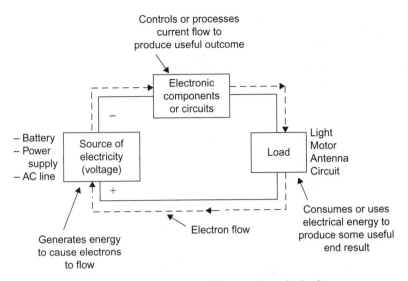

FIGURE 2.2   Basic electrical circuit used to describe every electronic circuit.

during a specific period of time. The charge or quantity of electrons (Q) is measured in coulombs (C). One coulomb of charge is equal to 6,242,000, 000,000,000,000 or 6.242×10^{18} electrons. If 1 coulomb moves past a point in 1 second, we say that the current is 1 ampere (A). One coulomb per second (1 C/s) equals 1 A.

Direction of Current Flow

The current flowing in electronic components and circuits consists of electrons that are influenced by a charge or voltage source. The early pioneers of electricity did not know that electricity was the movement of electrons. They knew that there was some kind of movement, of a charge or current, but they did not know what it was, or in which direction it flowed. In order to explain electrical phenomena and perform analyses of components and circuits, they simply assumed a direction of current flow that everyone agreed to and used. Their original assumption was that current flowed from positive to negative, the opposite of the actual direction. This has become known as conventional current flow.

Even after the discovery that real current was electrons moving from negative to positive, scientists and engineers continued to assume a positive-to-negative current flow. And science and engineering professors have continued to teach and support the conventional current flow concept. Most science and engineering textbooks illustrate and teach conventional current flow. Old traditions die hard, or maybe never die.

As it turns out, it does not really matter what direction current flows. Any analysis, design, or explanation can be successfully carried out using either electron flow or conventional current. One way is not any better than the other in

(Continued)

Direction of Current Flow (Continued)

describing or understanding electrical or electronic phenomena. Why academia has perpetuated the myth of conventional current flow rather than accepting and teaching the truth is not known. It no doubt stems from long-time heritage, custom, and convention.

Conductors, Insulators, and Semiconductors

A conductor is a material that has many electrons easily freed up by an external voltage. Current flows through conductors easily. They have what we call low resistance. Most good conductors are metals like copper, silver, or aluminum, with loosely bound electrons that can be freed by an external voltage to create current flow. Copper wire is the most commonly used conductor because of its low cost and its ability to be formed into many different shapes and sizes. Most metals are conductors but so is salt water.

Insulators are just the opposite of conductors. These are usually compounds in which the electrons are tightly bound together with the nuclei of the atoms. Even with lots of voltage applied, the electrons are hard to strip away to make current. Insulators keep current from flowing. Some common insulator materials are glass, ceramic, and plastics.

A third category of material is called *semiconductor*. It is a material that can be changed to make it a good conductor or a good insulator or anything in between. Semiconductors are used to make the transistors, diodes, and integrated circuits. The most common semiconductor is silicon. Others are germanium and carbon. Compounds such as gallium arsenide (GaAs), indium phosphide (InP), and silicon germanium (SiGe) are examples. Most integrated circuits (ICs) or chips are made of silicon.

MAGNETISM

You are probably asking, what does magnetism have to do with anything? As it turns out, it has lots to do with electricity and electronics. As you will see, the two are tied together significantly, meaning you need to know about both to get the big picture. Magnetism is essential to all electrical and electronic phenomena. Without magnetism, there would be no electricity or electronics.

Magnetic Fields

Magnetism is an invisible force that has the power to attract pieces of iron, or alloys of iron such as steel. Iron, steel, and the metals nickel and cobalt are called magnetic materials because they can be magnetized to generate a magnetic field, or they support the development and passage of a magnetic field. Magnets are pieces of such materials that exhibit this force. Magnetism has no effect on nonmagnetic materials such as other metals (aluminum, brass, etc.), or on objects made of wood, plastic, or glass.

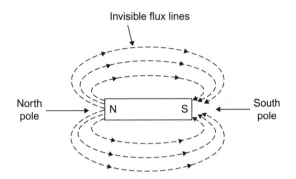

FIGURE 2.3 Permanent bar magnet illustrating the magnetic lines of flux that surround it.

You have no doubt experienced the effects of magnetism yourself. You may have played with a horseshoe magnet as a kid, or used bar magnets like the one in Figure 2.3 in a school lab. And you most likely have used flat magnets to hold paper notes on your refrigerator or file cabinet. These are called permanent magnets because they retain their magnetic force indefinitely.

A magnet is surrounded by an invisible magnetic force field made up of magnetic lines of force or flux lines. These flux lines flow out of one end of the magnet, the north (N) pole, and into the other end, the south (S) pole, as shown in Figure 2.3. The ends of the magnet are the points of the heaviest concentration of the magnetic field.

The basic law of magnetism follows: *opposite poles attract, and like poles repel.* When you get two permanent magnets together, the north and south poles will attract one another. On the other hand, two north poles will repel one another. Two south poles will also physically push away from one another. Permanent magnets are used in a variety of electrical and electronic equipment. Their most common use is in motors, where the attraction and repulsion of magnets create rotary motion.

Electromagnetism

Magnetism is also produced by electricity. Whenever electrons flow in a conductor, they produce a magnetic field. This effect is called *electromagnetism*. Figure 2.4A shows how the magnetic lines of force encircle a wire through which current is flowing. Note that the direction of the lines of force depends on the direction of the current flow.

The strength of the magnetic field around a wire depends on the magnitude of the current flowing. High current (composed of many electrons) produces a strong magnetic field. Despite the current amplitude, however, the magnetic field weakens because it spreads out quickly, even at a short distance from the wire.

If you make a coil out of the wire, as shown in Figure 2.4B, the lines of force around each turn are added together. The result is that a more powerful,

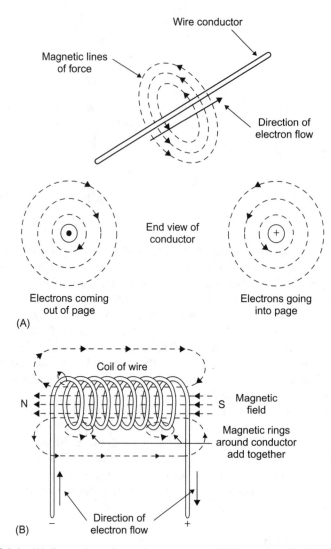

FIGURE 2.4 (A) Current in a wire produces a magnetic field around it. (B) Coiling the wire increases the strength of the magnetic field.

highly concentrated magnetic field is produced. In fact, the coil simulates a bar magnet with north and south poles.

As indicated earlier, the strength of the magnetic field depends on the current amplitude in the wire. High current produces a strong field. The magnetic field is also increased by coiling the wire, which helps concentrate the lines of force. The closer the turns of the wire, the stronger the field. You can also increase field strength for a given current level by simply adding more turns

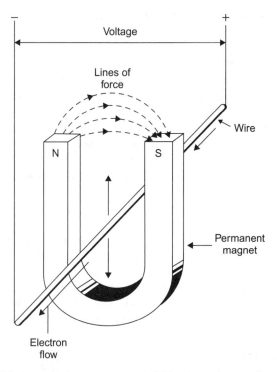

FIGURE 2.5   Relative motion between a magnetic field and a conductor such as a wire causes a voltage to be induced into the wire making it a voltage source.

to the coil. Strong electromagnets are made by tightly coiling together many turns of fine wire in several layers.

   Another way to increase the strength of the magnetic field is to insert a bar or core of magnetic material inside the coil. Because flux lines flow easier in iron or steel than in air, an iron or steel bar inside the coil helps concentrate the lines of force. Most of the flux lines flow through the core because it represents a lower path of resistance than the surrounding air for the lines of force. The result is a very strong field.

Electromagnetic Induction

Another important principle of magnetism is electromagnetic induction. This principle states that whenever there is relative motion between a conductor and a magnetic field, a voltage will be induced into the conductor.

   Refer to Figure 2.5. If the wire is moved up or down so that it cuts through the magnetic lines of force between the magnet poles, voltage will be induced in the wire and cause current to flow. Voltage will also be induced into the wire if the wire is held in a fixed location and the magnetic field is moved so that the flux lines cut across the wire. The wire actually becomes a voltage source. That voltage will cause current to flow. The direction of current flow in the wire depends on both the direction of the magnetic field and the direction of the relative motion between the magnetic field and the wire.

The amount of the induced voltage depends on the number of flux lines cut and the speed of cutting. The greater the number of lines of force, the higher the induced voltage and the greater the current flow. The highest induced voltage occurs when the conductor cuts across the flux lines at a right angle. If the wire moves parallel with the lines of force, no voltage is induced.

The induced voltage can also be increased by coiling the wire. More turns passing through the magnetic field will produce greater induced voltage and current flow. The motion in the magnetic field causes a small voltage to be induced into each turn of wire. All of the voltages will add together to produce a higher voltage.

In the examples above, we have assumed that the magnetic field comes from a permanent magnet. However, it doesn't really matter where the field comes from. An electromagnetic field works just as well. Electromagnetic induction is the principal method of generating electricity in the world. Most electrical power is produced by giant generators containing large coils of wire that rotate in a magnetic field. The generator or alternator in your car works the same way.

VOLTAGE SOURCES

To make current flow, we must have voltage. That voltage will produce either direct current (DC) or alternating current (AC). DC is electron flow in only one direction. AC is current that flows in one direction and then reverses and flows in the opposite direction.

DC Voltages

DC may be a fixed, steady current, one that varies, or one that switches off and on periodically. A DC voltage source is one that causes direct current to flow one way through the circuit, component, or conductor. It has positively and negatively charged terminals of a fixed polarity. It always causes electrons to flow from negative to positive.

Figure 2.6A shows a fixed, positive DC voltage source, which is usually either a battery or an electronic power supply. We call it a positive source because the negative terminal of the DC source is connected to a common point called the *ground*. The other terminal is positive with respect to ground. Voltages are always looked at or measured between two points with one point (ground) being the reference. The polarity of the voltage is the polarity of the terminal not connected to ground.

The remainder of Figure 2.6 shows graphs of DC voltage over time. In electronics we normally view DC or AC voltages with respect to time. The positive DC voltage is shown in Figure 2.6B as a horizontal dashed line.

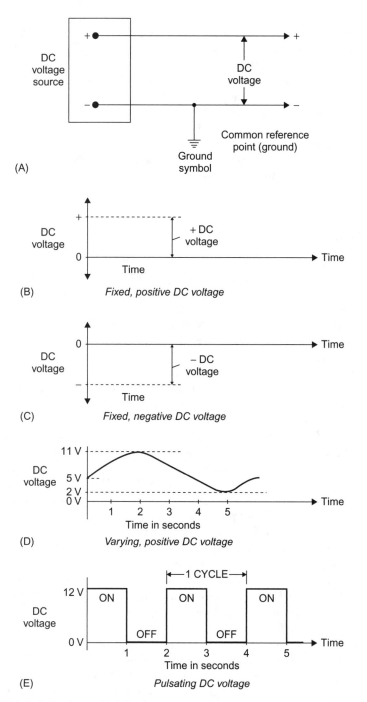

(A)

(B)   *Fixed, positive DC voltage*

(C)   *Fixed, negative DC voltage*

(D)   *Varying, positive DC voltage*

(E)   *Pulsating DC voltage*

FIGURE 2.6 DC voltages. **(A)** DC voltage source such as a battery or power supply. One side is usually grounded to serve as a reference. **(B)** Fixed, continuous positive DC voltage shown over time. **(C)** Negative DC voltage. **(D)** Varying positive DC voltage. **(E)** DC voltage pulses.

A negative DC voltage is shown in Figure 2.6C. The dashed line representing the voltage below the zero line indicates that it is negative. The voltage is negative with respect to the reference point known as ground.

Figure 2.6D shows a varying positive DC voltage. It begins at +5 volts, rises upward to +11 volts, and then drops to +2 volts over a period of 5 seconds. The left-hand scale indicates the voltage level.

Figure 2.6E shows a pulsating DC, that is, one that turns off and on. This voltage switches on to +12 volts, stays on for 1 second, and then switches to zero for 1 second. Then this on–off cycle repeats. One cycle is the time of one ON pulse and one OFF pulse or in this case 2 seconds. Such pulse signals are usually called *binary* or *digital signals*.

AC Voltages

As its name implies, AC current flows in two directions in a conductor, component, or circuit, but not at the same time. The electrons move in one direction, then reverse and flow in the other direction, both for a short period of time. The direction of flow reverses repeatedly.

The polarity of an AC voltage reverses periodically. During one period, the terminals of the voltage source are + and −, but later reverse to − and +. This reversal continues to repeat as the voltage amplitude varies. Since it is the voltage that causes current to flow, if the polarity of the voltage changes, then naturally the direction of the electron flow will also change periodically. The electrons just flow back and forth in the circuit.

An AC voltage source is usually represented by just a circle with a varying wave inside as illustrated in Figure 2.7A. It may be the AC outlet in your home, a signal generator, or a type of electronic circuit called an *oscillator*. AC voltages and currents can assume many shapes. The most common electrical and electronic AC voltage is a sine wave, as shown in Figure 2.7B. The voltage rises gradually from zero to a positive peak, and then drops to zero. Then it reverses polarity, rises to a negative peak, and returns to zero. Each rise and fall is called an alternation. One positive and one negative alternation form a cycle. The cycles repeat over and over. The number of cycles that occur per second is called the frequency.

The AC power-line voltage that comes into your house and appears at the outlets is a sine wave. Its frequency is 60 cycles per second (cps). The term *hertz* (abbreviated Hz) is normally used instead of cps to express frequency. The power-line frequency, therefore, is 60 Hz. All radio signals are sine waves, but have very high frequencies.

AC signals can have a variety of other shapes as well. Figure 2.7C shows a square or rectangular wave. It is made up of pulses of alternating polarity voltages that switch from +6 to −6 volts. Figure 2.7D shows a triangular wave, and Figure 2.7E, a complex waveform that might represent voice or video. Regardless of their shape, AC signals have one thing in common: they reverse the polarity and direction of current flow periodically. Their frequencies can range from a fraction of a cycle per second to billions of cycles per second.

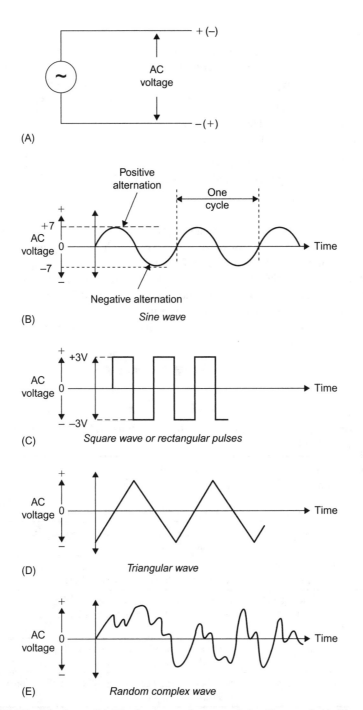

FIGURE 2.7 AC voltages. (**A**) AC voltage source drawn as a circle with a wave inside. The output polarity changes periodically. (**B**) Sine wave, the most common AC voltage shape. (**C**) Rectangular wave or AC pulses. (**D**) Triangular wave. (**E**) Random AC wave that may represent voice or video.

AC Frequency

Frequency is measured and expressed in units called cycles per second (cps) or Hz. Hertz is the more common unit. To express higher frequencies, the units kilohertz (kHz), megahertz (MHz), and gigahertz (GHz) are used.

1 kHz = 1000 Hz
1 MHz = 1 million Hz or 1000 kHz
1 GHz = 1 million Hz, 1 million kHz, or 1000 MHz

Table 2.1 presents a summary of various frequency ranges that you will encounter and the types of signals in those ranges.

AC Voltage Measurement

To measure or represent the amount of AC in a circuit, we usually state the peak, peak-to-peak, and root mean square (rms) values. The peak value of a wave is simply the voltage measured from the zero line to the maximum positive (or negative) value. In Figure 2.7B, the peak value of the sine wave (V_p) is 7 volts. The peak-to-peak value, normally abbreviated V_{pp}, is the positive peak added to the negative peak. Or, since the positive and negative peaks are identical in a sine wave, the peak-to-peak value is two times the peak value, or in this case 14 volts.

$$V_{pp} = 2V_p$$

TABLE 2.1 The Frequency Ranges of the Electromagnetic Spectrum and Their Applications

Frequency Range	Types of Signals
20 Hz to 20 kHz	Audio, sound, voice, and music.
300 kHz to 3 MHz (LF)	Low radio frequencies. AM broadcast.
3 MHz to 30 MHz (HF)	High-frequency radio, shortwaves. Broadcasting, amateur radio, CB.
30 MHz to 300 MHz (VHF)	Very-high radio frequencies. 2-way radio, mobile, marine, aircraft; TV and FM broadcast.
300 MHz to 3 GHz (UHF) Anything above 1 GHz is called microwave.	Ultra-high radio frequency. 2-way radio, TV broadcast, cell phones, GPS satellites, wireless broadband, wireless networks.
3 GHz to 30 GHz (EHF)	Extremely high radio frequencies. Radar, satellites, network radios.
30 GHz to 300 GHz	Super-high radio frequencies. Radar, satellites, network radios.

Another way to measure AC voltage is to use an average value known as the root mean square (rms), or effective value. The rms value is computed by multiplying the peak value by 0.707, or:

$$V_{rms} = 0.707 \, V_p$$

Thus, the rms value of the sine wave is $0.707 \times 7 = 4.949$ volts. Most sine wave voltages are measured and stated in terms of the rms value. The amount of AC voltage available at a standard AC outlet, for example, is 120 volts rms. The rms value as discussed above is applied only to sine waves. The significance of the rms value is discussed later when we address power.

PRACTICAL DC AND AC VOLTAGE SOURCES

Just what is the source of DC and AC? Such voltages originate from a variety of sources.

DC Voltage Sources

Batteries

The most common source of DC voltage is a battery. A battery is a collection of cells. A cell is the basic unit of voltage generation that creates electricity by chemical action. A battery is made up of two or more cells connected together. Figure 2.8 shows the configuration of a cell. It consists of two metal electrodes immersed in a solution called an electrolyte. The chemical interaction between the metal electrodes and the electrolyte produces a separation of charges. This separation of charges results in an excess of electrons on one electrode (the − terminal of the cell) and a shortage of electrons on the other electrode (the + terminal of the cell). If a conductor is connected externally between the terminals, electrons will flow from − to +.

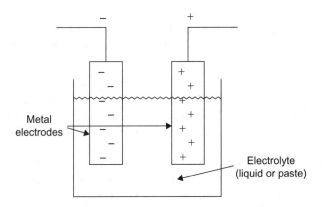

FIGURE 2.8 Basic concept of any battery.

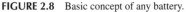

TABLE 2.2 Most Popular Types of Cells/Batteries and Typical Applications

Materials	Voltage	Type	Applications
Alkaline	1.5	Primary	AAA, AA, C, D cells; flashlights; toys; and much more
Mercuric oxide	1.35	Primary	Watches; calculators
Silver oxide	1.6	Primary	Hearing aids; watches
Nickel-cadmium	1.2	Secondary	Tools
Nickel metal hydride	1.2	Secondary	Laptops; cell phones
Lead acid	2.1	Secondary	Cars; golf carts
Lithium ion	4.1	Secondary	Cell phones; iPods; laptops

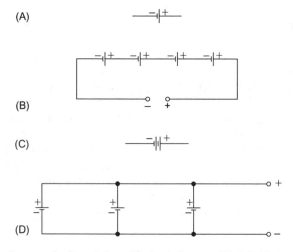

FIGURE 2.9 Battery and cell symbols used in circuit diagrams. (**A**) Cell. (**B**) Four cells in series making a battery. Cell voltage add-up. (**C**) Symbol for battery of any number of cells. (**D**) Cells in parallel to increase current capability.

The amount of voltage produced by the cell depends on the type of electrodes and the electrolyte. Various chemical combinations generate different voltages. Table 2.2 lists the most commonly used cells and the voltage they generate.

The schematic symbol of a cell is illustrated in Figure 2.9A. To create a battery, two or more cells are connected together. The cells are usually connected in series form, that is, end to end and − to +, as shown in Figure 2.9B. The cell voltages add up to produce the total battery voltage. For example, six small alkaline cells are connected in series to create a $1.5 \times 6 = 9$-volt transistor-radio battery. A car battery consists of six lead-acid cells connected in series; each produces 2.1 volts for a total battery voltage of $6 \times 2.1 = 12.6$ volts. If four nickel-metal-hydride cells are connected in series, as in Figure 2.9B, the total battery voltage is 4.8 volts.

The simplified schematic symbol for a battery is shown in Figure 2.9C. Instead of drawing all of the individual cells, the simplified battery symbol is used.

The cells listed in Table 2.2 can be separated into two categories, primary and secondary. *Primary cells* cannot be recharged. As the chemical action produces voltage, the electrodes and electrolyte are actually used up. At some point, no further chemical action is produced, and no voltage is generated. The cell must be discarded and replaced.

Secondary cells, on the other hand, can be recharged. In these cells, the chemical action can be reversed by connecting an external DC voltage to the battery. This is called *charging*. Charging forces electrons into the battery, producing a reverse chemical action that rejuvenates the electrodes and the electrolyte. The cell can be discharged and charged repeatedly, giving it a very long life.

The amount of current that can be drawn from a cell depends on its size and the quantity of the materials used. Large electrodes and electrolytes can produce more current, but the voltage stays the same; the type of material, not its volume, determines voltage.

For example, a large D-size flashlight cell and a tiny AAA penlight cell are both made of alkaline material so their output voltages are both 1.5 volts. The D-cell, however, can produce more current and will have a longer life because it is larger.

A greater current capacity can be obtained by simply connecting two or more cells in parallel. In a parallel connection, the cells are wired across one another, all + leads connected together, and all − leads connected together, as shown in Figure 2.9D. The output voltage remains the same, but the current output can be higher.

Solar Cells

A special type of voltage-producing cell is the so-called *solar cell*. Also called a *photovoltaic* (PV) cell, a solar cell converts light into electrical energy. When the sun or artificial light shines on the cell surface, a voltage is generated. The most commonly used material is silicon, and a typical cell generates about 0.5 volt. Other materials, such as cadmium sulfide and gallium arsenide, are also used.

You have probably seen solar cells used in hand-held calculators. Big arrays of solar cells are now being adopted as an energy source for homes and businesses. Another important application is spacecraft. Virtually every satellite, deep-space probe, or other space vehicle, uses solar cells for power. Huge panels of solar cells connected in a complex series–parallel array produce a powerful battery that can operate all electrical and electronic equipment on board. Typically, however, such a solar battery is primarily used to recharge the main spacecraft power source, which may be a nickel-cadmium secondary battery, as shown in Figure 2.10. The voltage produced by this array is 2 volts.

Fuel Cell

A fuel cell is a special type of battery that combines oxygen and hydrogen gas to produce voltage and electron flow. That means you have to have a supply of hydrogen gas, which is usually stored in a bottle or other container. Typically, the oxygen can be extracted from the air, but if not you will need a bottle of oxygen as well. When you combine the two, charges separate and plus and minus terminals are formed. Most fuel cells generate between 0.5 volt and 0.9 volt, so you have to put many of them in series to be useful. Usually they are bulky and inconvenient because of the need for the gases. But as long as oxygen and hydrogen are supplied, the fuel cell produces the voltage. Heat and water are by-products of the chemical process, making them essentially impractical for many applications.

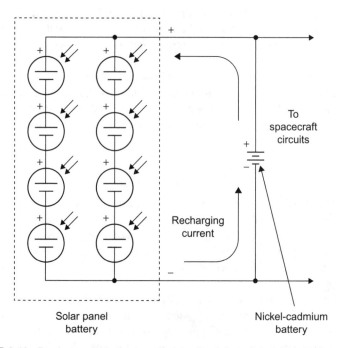

FIGURE 2.10 Panel comprised of solar cells in series and parallel make up DC source that charges the battery in a spacecraft.

Power Supplies

Electronic circuits require a source of DC for proper operation. A great many electronic products are still powered by batteries because they are small and portable, such as cell phones, iPods, and the like. Other electronic equipment is designed to operate from the AC power line. Such equipment contains an electronic power supply that converts AC into the required DC voltages. Computers, TV sets, and other nonportable electronic equipment have an AC power supply.

AC Voltage Sources

Generators

The AC that appears at the outlets of our homes and offices is produced by huge electromechanical generators owned and operated by an electrical utility company. A source of mechanical power—such as water, or a steam turbine powered by oil, coal, or nuclear fuel—operates the generator. The turbine converts steam into mechanical energy to turn the generator. The generator rotates coils in a strong magnetic field to produce a voltage. The speed of rotation of the coils determines the frequency. That voltage is a 60-Hz sine wave.

The big wind generators popping up across the landscape also produce AC sine waves, but because the rotational speed depends on the wind speed, the AC frequency will vary, which is unacceptable. Therefore, the AC voltage they produce is converted to DC, and then back into AC of the correct frequency (60 Hz) to be used or put on the AC power grid.

Another example is the alternator in your car. "Alternator" is just another word for an AC generator. The alternator, driven by the car engine through a belt, produces AC output that is converted to DC by diodes making up a rectifier. The rectifier output is DC to charge the battery.

Inverters

Electronic circuits called inverters are used to produce AC from a DC source. This is common in solar power systems. The solar cells produce DC, which keeps batteries charged. The DC is then converted to AC by the inverter.

In electronic equipment, AC signals are generated by circuits called *oscillators* or *function generators*. The AC signals are sine waves, square (rectangular) waves, or triangular waves, among others.

Analog and Digital Signals

You have no doubt heard the terms "analog" and "digital." These terms distinguish between the two main types of voltages or signals in electronic equipment. Analog signals are voltages that vary smoothly and continuously over time. You saw some in previous figures. Analog signals may be either DC or

AC. A continuous DC voltage is analog, as well as a varying DC signal. A sine wave and the random signal in Figure 2.7E are also analog signals.

Analog signals come from microphones, video cameras, and sensors that detect light level, temperature, and other physical conditions. The most common types of analog signals are voice or music from a microphone and video from a camera.

Digital signals are pulses or on–off signals. The DC pulses in Figure 2.6E and 2.7C are digital. Most digital signals are binary in nature, meaning they only have two voltage values. In Figure 2.6E the two values of voltage are $+12$ and zero volts. In Figure 2.7C the two voltage levels are $+3$ and -3 volts.

You will see both analog and digital signals in all sectors of electronics. Today, most signals are digital. Special circuits called analog-to-digital converters (ADC) and digital-to-analog converters (DAC) are used to translate between the two types of signals.

Fourier Theory and the Frequency Domain

We normally think of electronic signals as voltages that vary over time. Waveforms showing signals are said to be in the time domain. Time is the horizontal axis in a graph. But there is another way to look at signals in the frequency domain where frequency is on the horizontal axis. For example, a sine wave is shown as a single vertical line at its frequency as Figure 2.11A shows. The height of the line is representative of the voltage level, either peak or rms.

The big question is, how do you show a complex signal like a square wave in the frequency domain? The only way to do this is to use the Fourier theory that says that any complex nonsinusoidal wave is made up of a group of sine waves added together. There is a fundamental sine wave that occurs at the signal frequency plus harmonic sine waves. A harmonic is a sine wave whose frequency is an integer (whole number) multiple of the fundamental sine wave. The second harmonic of a 50-MHz sine wave is a 100-MHz sine wave. The third harmonic is a 150-MHz sine wave and so on. If you add the fundamental sine wave to multiple harmonics of differing amplitudes and phases, you can create any shape of wave.

For instance, a square wave is made up of a fundamental sine wave and all the odd harmonics. A 1-MHz square wave is made up of a fundamental 1-MHz sine wave and harmonics of 3 MHz, 5 MHz, 7 MHz, and so on. According to Fourier theory, it takes an infinite number of harmonics but in reality only a few (like up to the fifth or seventh) are needed to approximate the square wave. The frequency domain plot of a square wave therefore would look like the one in Figure 2.11B.

When looking at a time domain view of a signal, try to keep a frequency-domain picture in mind at the same time. You cannot actually see the third or fifth harmonics in the time domain plot but they are buried in there just the same.

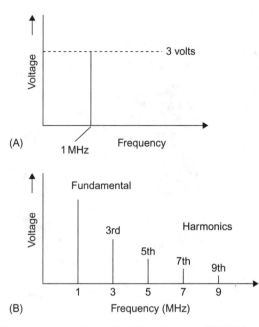

FIGURE 2.11   Frequency-domain views. **(A)** 1-MHz sine wave. **(B)** 1-MHz square wave.

Project 2.1

Getting Familiar with a Multimeter

One of the things you are going to need as you work with electronic parts, circuits, and equipment is a multimeter. This is a test instrument that measures voltage, current, and resistance. Some also measure capacitance and other things. These multimeters are usually hand held and battery operated. There are two basic types, analog and digital. Analog meters use a dial to show the measured value. Digital multimeters (DMMs) have a liquid crystal display (LCD) to show the measurement. See Figure 2.12. The two types work equally well; which one you use is typically a matter of preference for meters versus digital displays. I recommend a digital unit.

The good news is that multimeters are not expensive. You can get one for as little as $10, but a better unit may cost you several times that. The cheaper units are kits that you will need to build yourself. I do not recommend a kit if you are just starting out. Later in this book I will discuss kits in more detail. In any case, you need to buy a multimeter and learn to use it. Plan to buy one as soon as you can. Check your local Radio Shack store or buy online.

Three good sources of multimeters by mail order or web order follow:

Elenco Electronics: www.elenco.com

Jameco: www.jameco.com

Kelvin Electronics: www.kelvin.com

Ask for a printed catalog when you inquire.

(Continued)

Project 2.1 (Continued)

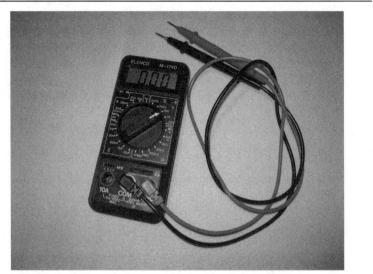

FIGURE 2.12 A low cost digital multimeter that measures DC and AC voltages, current, and resistance.

Project 2.2

Hardware Sources

As an electronic experimenter, you should be familiar with buying parts and equipment. Radio Shack stores still stock some parts but today most hobbyists and experimenters buy via the mail or online. The most popular sources of parts and equipment are All Electronics, Digi-Key, and Jameco. And you will discover others. You can go to their websites, but you really should get on their mailing list and get a catalog. Do that as soon as you can. Links to these sources follow. You can get your multimeter from one of them as well.

All Electronics: www.allelectronics.com

Digi-Key: http://www.digikey.com

Jameco: http://www.jameco.com

Ramsey Electronics: http://www.ramseyelectronics.com

Project 2.3

Magazines of Interest

As you are learning electronics, you need to be continuously aware of what is going on in the industry and with the technology. Magazines are a good source of that information. One that you should read on a regular basis is *Nuts & Volts*. It is

(Continued)

Project 2.3 (Continued)

available at some newsstands and bookstores, but if you cannot locate it, go to the website and subscribe. This magazine has lots of articles and columns, including construction projects. It is a great way to stay up to date and learn more. *Nuts & Volts* also publishes *Servo* magazine, which is aimed at people who like to build and play around with robots.

Another recommended magazine that has an embedded controller emphasis is *Circuit Cellar*. This publication contains great educational articles and projects. Get a subscription if you are serious about learning more.

Another interesting magazine is called *Make*—it is not strictly electronics but it usually contains some good electronic projects. It also features other mechanical projects such as robots, among other things. This one is definitely worth a look the next time you are at the newsstand.

Nuts & Volts and *Servo*: www.nutsvolts.com

Circuit Cellar: www.circuitcellar.com

Make: www.makezine.com

Project 2.4

Measuring Voltages

A good first step in using a multimeter is to measure voltages. Batteries are a good place to start. Look around the house and locate some batteries that you commonly use. Take them out of the equipment before you measure them. You will most likely find AA or C cells. Set the multimeter to measure DC according to the meter user's manual and connect the test leads to the battery. The red lead is +, the black lead is −. The LCD on the meter should show the voltage. It should be 1.5 volts. You can also check batteries in some cordless phones, cell phones, or other devices. You may be able to remove the battery from a laptop. In each case, check to determine the battery technology—lithium ion, alkaline, and so on.

Two key points to remember. First, if you get the test leads reversed, the red to the − terminal of the battery and the black to the + terminal, the meter will read a negative value. That just means that the leads are reversed, nothing more. Second, you cannot typically judge the condition of the battery by measuring its voltage. With no load on the battery, no current is drawn from it so you are just measuring the characteristic voltage. If you connect the battery to something, current will flow and its voltage will decrease slightly. If it decreases a lot, the battery is going bad. If the cell or battery voltage is several tenths of a volt lower than the characteristic value given in Table 2.2, the battery is probably bad.

You could also go out to your car and measure its battery voltage. Be careful. You cannot get shocked but you could short something. When the engine is off, the battery will probably register about 12 volts. With the engine running, the battery voltage will be a bit higher, as much as 15 volts or so. The generator is charging the battery, making its voltage a bit higher.

(Continued)

Project 2.4 (Continued)

One last thing: you can use the multimeter to measure the AC line voltage at an outlet. Only do this if you have probes that will fit into the slots on the outlet. **And be super careful as you can kill yourself. Be paranoid, even.**

Set the multimeter for AC voltage at an appropriate level of at least 120 volts, according to the multimeter manual. Place the probe tips into the AC outlet and read the voltage.

Do NOT touch any metal with the probe tips!!

Your value will depend on things like how far you are from the power station, what appliances and other items are on, and time of day. In any case, the voltage range is about 95 to 130 volts. It is rarely lower than 110 or higher than 125 volts. The typical value is 120 volts, or what is called *nominal*. I measured mine at 117 volts.

Chapter 3

The Systems versus Components View of Electronics

A Fresh New Way to Learn about Electronics

In this Chapter:
- Systems defined.
- Examples of systems.
- Introduction to basic electronic components.

INTRODUCTION

When learning about electronics in the past, you started with basics of physics and math, and then went on to learn electrical theories. Next came instruction in how electronic components work, and then you combined them into circuits. Then you put the circuits together into specific pieces of equipment. Finally, at last, knowing the equipment, you could understand the application, such as TV or computers or audio or home system control or whatever. This approach to learning is called the bottom-up approach. You learned the details and added them all together to get to what you really wanted to know about—the application and its technology.

That is still the way you learn electronics in college today, and it is the approach taken by most other electronics books you may have seen. This book doesn't take that old legacy path. Instead it skips most of the early background and theory, goes right for details on the circuits, and then builds to the applications. It is a much more interesting way to learn electronics. And it is also far more relevant to how electronic equipment is implemented today. We call this way of learning the "systems view" of electronics.

WHY SYSTEMS AND NOT CIRCUITS?

What is the advantage of this approach? First, thanks to modern semiconductor technology, most electronic components and circuits are inside chips or integrated circuits (ICs). You can't see them, and you can't access, test, or replace

doi: 10.1016/B978-1-85617-700-9.00003-5

them. About all you can do is learn what they do, and then work with the various input and output signals. Since you do not have access to the components and circuits, there is some argument as to why you need to learn all that circuit theory detail. If you are an engineer designing ICs, you definitely need to know this. Otherwise, it is just a waste of time unless you want to know just for the heck of it. Some of it is nice to know, of course, but in most cases it is not necessary. What you really need to know is what the circuit is, what it does, what its characteristics are, and what the inputs and outputs should look like.

A good example is the power supply in a PC. If it goes bad, you just replace it with a $30 unit that you can buy online. Troubleshooting the power supply would take many hours and require some expensive test equipment and access to spare parts. The whole repair could take hours and at a typical rate of, say, $50/hour, or more, it could cost hundreds of dollars and take weeks to get the new parts. A quick replacement with the $30 unit takes only about a half hour at most. So for less time and money, you get your PC working faster. Not a bad deal at all.

A cell phone is the same. It takes hours of troubleshooting and very expensive test equipment to troubleshoot and repair one. A typical repair would cost hundreds of dollars. Consequently, it is rarely done. It is cheaper to get a new cell phone, and it will be a better one at that. And it may even be free if you renew your service contract. That's electronics today.

The ability to analyze circuit operation by tracing electron flow through the various components used to be necessary. That made it necessary to know about all the components in detail and how they all come together to make a circuit. But with the systems view, you rarely need this knowledge unless you are an engineer designing the circuits or the equipment. In most cases you are not, so why invest the time?

The systems view is a better approach, as you do learn about the circuits and the equipment at a higher level. Later, if you want or need to, you can delve deeper into the circuit and component details, just for fun or to satisfy your curiosity. We won't do that in this book. But you will get the system view and big picture that will serve you well in learning electronics.

The systems view is more important today because ICs have made it possible to largely ignore the internal details. The ICs and systems on a chip (SoC) have resulted in the throw-away electronics society in which we live. It is no longer possible or economical to repair most electronic equipment thanks to ICs. The cost of labor, the cost of test instruments to troubleshoot equipment, and the difficulty of replacing ICs has made it more practical to just discard a bad piece of electronic equipment rather than repair it. It costs less and you get a newer and better replacement product as a result. Not a bad deal.

Yes, it is still possible to repair some equipment. For example, big expensive equipment such as a magnetic resonance imaging (MRI) machine is usually repairable, but it is done by replacing not bad components, but by replacing defective modules, subassemblies, or printed circuit boards.

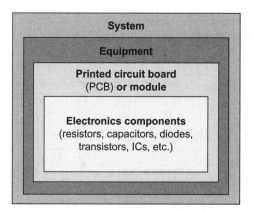

FIGURE 3.1   Hierarchy of electronics from systems to components.

You don't have to know how to analyze circuits to do that. You just trouble-shoot by observing signal flow, measuring inputs and outputs, and replacing any module or subassembly that does not work. You get the equipment work-ing sooner and at lower cost.

A good way to picture the systems view is shown in Figure 3.1. It all begins with the components that make up the circuits that in turn make up the modules and subassemblies, usually on printed circuit boards (PCBs). Those are put together to form pieces of equipment. Finally, multiple pieces of equip-ment may end up being interconnected to form some kind of system. A few system examples will show you the idea.

A personal computer (PC) is a great example of a system (see Figure 3.2). It is made up of modules including a mother board with its processor, mem-ory ICs, and input/output (I/O) interfaces. There is a power supply module, a hard disk drive, a CD/DVD drive, and perhaps a video card. Then the system is completed when it is connected to the video monitor, keyboard, printer, and Internet modem.

A consumer home entertainment system is made up of a big screen HDTV set, a cable or satellite box, a DVD/Blu-ray player, and an audio amplifier/receiver with speakers. Other equipment may be involved. The TV itself or any of the other components are systems in their own right.

An iPod or MP3 player is a simple system. Shown in Figure 3.3, it consists of a memory, either a tiny hard disk drive or a solid-state flash memory that stores the music, an audio codec IC that does the digital music decompres-sion, digital-to-analog conversion and analog-to-digital conversion for storing music, a microcontroller that controls the whole thing including the selection controls, and the LCD readout. A pair of audio amplifiers and a headset com-plete the system. A power management system charges the battery and doles out DC to power the various ICs.

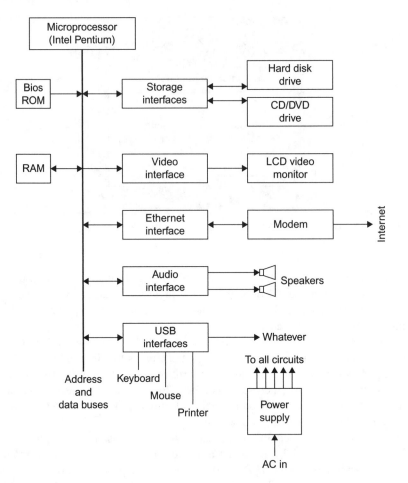

FIGURE 3.2   System block diagram of typical personal computer.

An MRI machine is a whole system. It is made up of the central bed surrounded by a superconducting magnet and radio frequency–sensing coils, high-power radio frequency (RF) amplifiers and detectors, the video imaging section, and a computer that runs the whole thing.

Anyway, you get the picture. There are big systems, little systems, and lots of in-between systems. Their main characteristic today is that you cannot get at the individual components and circuits, but there are exceptions.

In this book we look mainly at the circuits, equipment, and systems, not the components. We will give you a quickie introduction to the main components but move on to the circuits and systems.

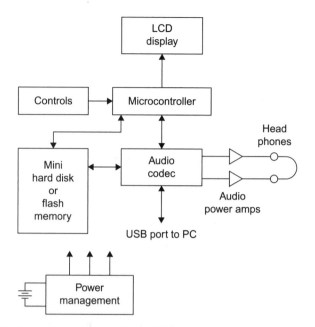

FIGURE 3.3   System block diagram of iPod or MP3 music player.

ELECTRONIC COMPONENTS

Most systems today are collections of ICs wired together on printed circuit boards and packaged in a housing of some sort. The MP3 player/iPod in Figure 3.3 is a great example. Each block in that diagram is an IC. In the past, systems were made with PCBs and other assemblies, but the circuits were made with discrete components, individual resistors, capacitors, diodes, transistors, and even some ICs. You will still see these here and there in modern equipment; just remember that most circuits and equipment are just a batch of ICs.

Just so you won't be ignorant of components, here is a nutshell introduction that should serve you well when learning electronics the system way. If you want to dig deeper, you can find many other books that will make your eyes glaze over with detail.

Switches

There is no simpler electrical component than a switch. It is just two (sometimes more) metal contacts and a mechanical device to open and close them. In one position the switch contacts are open and not touching so that no current will flow through them; the switch is "off." Moving the switch lever, the contacts are made to touch making a path for current to flow; the switch is said to be "on."

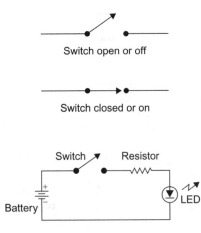

FIGURE 3.4   Switch and how it is used.

Figure 3.4 shows a switch in a circuit diagram called a *schematic*. This shows the switch connected to a battery and a light-emitting diode (LED) in a simple flashlight. The other component is a resistor to limit the current.

Resistors

Resistors are one of the most common components in electronics. These devices offer opposition to current flow; they limit or set the level of current in a circuit. The opposition to current flow is measured in ohms. The higher the ohms value, the more resistance it offers to electron flow.

Ohm's Law

There is one key electrical relationship you should know, called Ohm's law. It is simple and easy to understand. It says: The current in a circuit (I) is directly proportional to the voltage (V) and inversely to the resistance (R). It is usually expressed as a simple formula:

$$I = V/R$$

If you increase the voltage in a given resistance, the current increases. That makes sense since the voltage charge simply attracts and repels more electrons. Increasing the resistance makes the current decrease. Since resistance opposes current flow, more resistance with the same voltage will make the current decrease.

Keep this simple relationship in mind, as it is the heart of every electrical and electronic application.

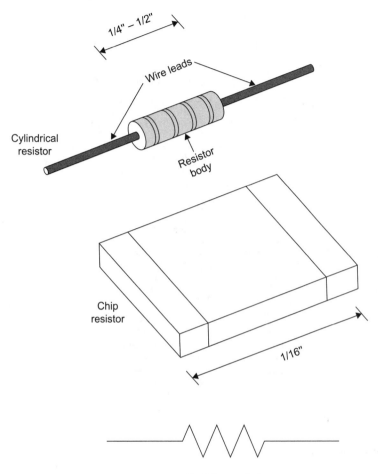

FIGURE 3.5   Types of resistors and schematic symbol.

Today most resistors are made along with all the other circuitry in an IC. But discrete component resistors are also still fairly widely used. The most common ones are called chip resistors and look something like the one shown in Figure 3.5. They are small and you can just make them out when looking at a printed circuit board. Older but still used resistors are cylindrical like the one shown in Figure 3.5. The schematic symbol used to represent a resistor in a schematic diagram is shown in Figure 3.5. Color-coded resistors use a set of color bands with a special code to indicate their value (see Figure 3.6).

An example of a common resistor circuit is the voltage divider shown in Figure 3.7. It is used to attenuate a voltage or signal or make it lower. The two resistors of the voltage divider are connected in series. When current flows, a

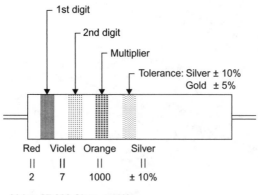

Value: 27,000 Ohms ± 10%

Resistor color code

Color of band	Value	Multiplier
Black	0	1
Brown	1	10
Red	2	100
Orange	3	1000
Yellow	4	10,000
Green	5	100,000
Blue	6	1,000,000
Violet	7	10,000,000
Gray	8	100,000,000
White	9	1,000,000,000

FIGURE 3.6 Standard resistor color code.

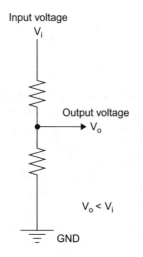

FIGURE 3.7 Voltage divider, the most common resistor circuit.

voltage develops across the two resistors. The input voltage is divided between the two resistors according to their value. In any case, the output is taken from one of the resistors and it is lower in value than the input.

Capacitors

A capacitor is a device that stores an electrical charge. It is made up of two metallic plates separated by an insulator such as plastic or ceramic or even air (see Figure 3.8). The symbol for a capacitor in schematics is also shown.

The basic function of a capacitor is to be charged or discharged. You charge the capacitor by applying a voltage to it as shown in Figure 3.9A. The electrons pile

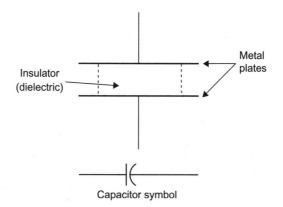

FIGURE 3.8   Capacitor construction and schematic symbol.

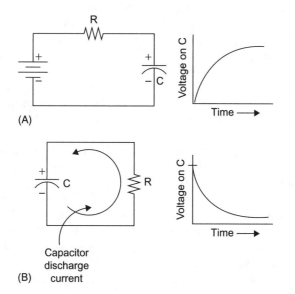

FIGURE 3.9   **(A)** Charging capacitor. **(B)** Discharging capacitor.

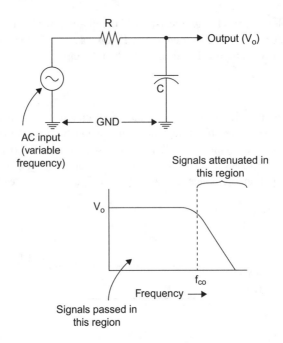

FIGURE 3.10 Low-pass filter, a common capacitor use.

up on one plate making it negative and are pulled away from the other plate making it positive. It takes a finite amount of time for the capacitor to charge as determined by the values of capacitance (C), stated in farads (F), and resistance (R). That time is called the time constant (T). In the charged state, the capacitor stores the voltage. The capacitor in its charged state is almost like a battery, as it can supply DC voltage to a circuit or load.

The energy stored in the capacitor is used by discharging it into a load such as a resistor (see Figure 3.9B). The electrons then flow from the negative plate through the resistor to the positive plate. Current flows until the charge is dissipated. Again, it takes time for the discharge to occur.

When capacitors are used with AC signals, they offer opposition to current flow called *capacitive reactance*. It is like resistance in that it determines current flow, but it varies with the frequency of the AC and capacitor value.

Capacitors perform all sorts of functions in electronic circuits. They are used for filtering and they are used to pass AC but block DC. Figure 3.10 shows a low-pass filter comprised of a resistor and a capacitor. This is a frequency-sensitive circuit that passes AC signals below a specific frequency called the *cut-off frequency* (f_{co}) and attenuates the frequencies above the cut-off. Figure 3.11 shows how a capacitor passes AC but blocks DC.

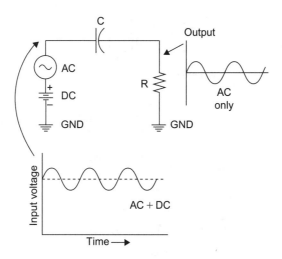

FIGURE 3.11 How capacitor passes AC sine wave but blocks DC voltage.

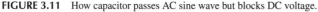

A Word about Ground

In schematic diagrams like Figure 3.10, you will see the triangular-shaped symbol labeled GND. This means ground. Ground is an electronic term that refers to a common connecting point. It may be a common wire, a metal chassis, a copper area on a printed circuit board, or some other conductive part or area. In this figure, it means that the bottom of the AC generator and the bottom of the capacitor are connected together.

Ground is also the common reference point for measuring all voltages.

Inductors

Inductors are mainly just coils of wire. When current flows through them, they produce a magnetic field and induce a voltage in themselves. This is called *self-induction*. The induced voltage has an opposite polarity of the applied voltage. The result is that the induced voltage causes opposition to the current flow. We often use that characteristic to control current flow in AC circuits. Like the capacitor, the inductor is an opposition to AC called *inductive reactance*. Figure 3.12 shows the symbol of an inductor as used in schematic diagrams.

Transformers

A transformer is usually comprised of two coils or windings on a common core. The schematic symbol is shown in Figure 3.13. A voltage applied to one winding called the *primary* causes current to flow and a magnetic field to develop. This magnetic field spreads out and induces voltage in the other

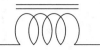

FIGURE 3.12   Schematic symbol for inductor.

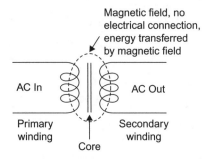

FIGURE 3.13   Schematic symbol for transformer.

winding called the *secondary*, thereby transferring electrical energy for one coil to another by way of the magnetic field. Transformers are used to step up or step down voltage levels of AC signals. They are also used for signal isolation as well as impedance matching.

Diodes

A diode is a semiconductor component that lets current flow through it in one direction but blocks current in the other direction depending on the polarity of the voltage applied to it. It acts like a polarity-sensitive switch. Figure 3.14 shows a typical diode, the diode schematic symbol, and how it is biased for current or no current.

Diodes are used mainly for rectification, the process of converting AC into DC. Figure 3.15 shows how a diode converts an AC sine wave into DC pulses. If you put a capacitor across the load resistor, it will charge up to the peak sine voltage and store it. The result is that the output is a near-constant DC value. Most electronic power supplies work like this.

Special diodes are made to emit light (LED), regulate voltage (zener diode), act as a variable capacitor (varactor), or perform as a switch (PIN diode).

Transistors

A transistor is a three-terminal semiconductor device that uses a small input signal to control a much larger output signal. One type of transistor is the bipolar junction transistor (BJT) shown in Figure 3.16. A small current applied to the base element is used to control a larger current flowing from emitter to collector.

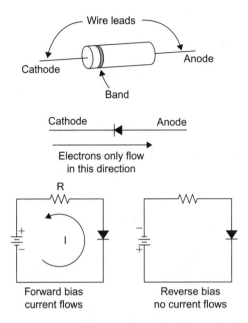

FIGURE 3.14   Diode, its schematic symbol, and how to bias it for conduction or cut-off.

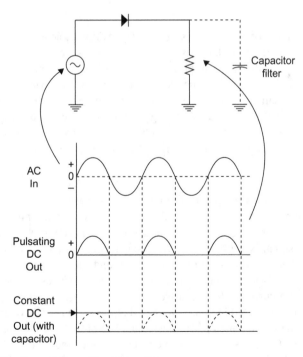

FIGURE 3.15   How a diode rectifies AC into DC.

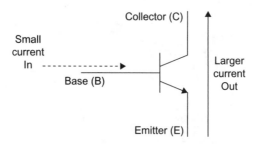

FIGURE 3.16 Schematic symbol for bipolar junction transistor (BJT) and its inputs and outputs.

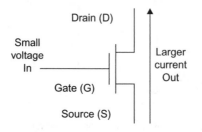

FIGURE 3.17 Schematic symbol for MOSFET and its input and outputs.

Another type of transistor is the metal oxide semiconductor-field-effect transistor (MOSFET). It too has three elements, as shown in Figure 3.17. A small voltage on the gate controls a larger current flowing from source to drain.

Transistors have two basic functions: amplification and switching. In amplification, the small input variation causes larger output current variation. The small input signal isn't really made bigger, but the larger output variation has the same shape and characteristics, resulting in the same effect. The transistor simply varies the larger DC current from the supply and shapes it like the input.

Amplifiers

Figure 3.18 shows a simple BJT amplifier. The base resistor sets the base current turning the transistor on so that current flows from emitter to collector. Its output voltage is a DC value at about half the supply voltage. Now, if a small AC signal is applied through the capacitor as shown, the base current will vary. This will vary the collector current and the voltage across the collector resistor. A small variation in the base current produces a larger variation in the collector current. The output voltage is an enlarged version of the input, except that it is reversed in phase. We say that the output is 180 degrees out of phase with the input, called *phase inversion*. The output is a DC signal, but to convert it to AC we pass it through a capacitor to the load. All amplifiers generally work like this.

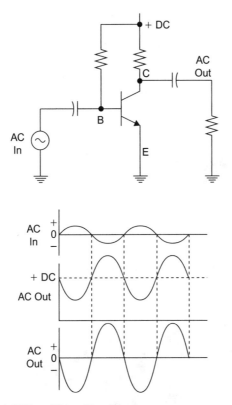

FIGURE 3.18   Simple BJT amplifier and how it works.

Switches

As a switch, the transistor output current is turned off or on by the smaller input current or voltage. A small input signal can switch a larger output current off and on. Switching transistors can switch very fast (in less than nanoseconds). Figure 3.19 shows a MOSFET switch. If the input voltage on the gate is zero, the MOSFET does not conduct. It acts like an open or off switch. The output then is just the supply voltage ($+V$) as seen through the resistor.

Now, if we apply voltage to the gate above a certain threshold level, the MOSFET will conduct. It acts like a closed or on switch so current flows in the resistor. The output is a very small value near zero. Figure 3.19 also shows what the inputs and outputs look like with a square wave applied.

Integrated Circuits

An IC is also called a chip. It is a complete circuit made on a single chip of silicon. All the components are made at the same time and interconnected to form an amplifier, microcomputer, memory, oscillator, radio transceiver, or

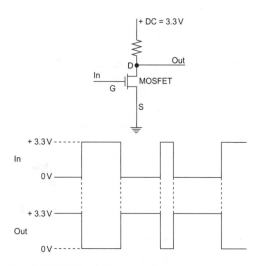

FIGURE 3.19   Simple MOSFET switch and how it works.

FIGURE 3.20   Typical integrated circuit today. This one is a dual-satellite TV tuner for set-top boxes. The size is only 7 × 7 mm. The pins are soldered to a printed circuit board. *(Courtesy NXP Semiconductor.)*

whatever the function. The IC is then connected to DC power and it works. All you do is apply the inputs, and new outputs are produced according to what the IC function is. Most circuits today are ICs.

There are all sizes and types of ICs. There are linear circuits for analog signals, digital circuits for binary signals, and mixed signal IC for both analog and digital signals. There are small ICs that are just an amplifier or logic circuit. Then there are even larger circuits like a computer memory or a microprocessor that is the heart of every computer. Finally, there are complete systems on a chip (SoC). Each year, it is possible to make the components on a silicon chip smaller so you can put more on a chip economically. That is why we have iPods, cell phones, and other really small devices that in the past would have been huge pieces of equipment. Figure 3.20 is representative of ICs today.

There are lots of other electronic components but these are the main ones. A few others are introduced in later chapters.

While this book attempts to get you thinking about the big picture—the systems rather than the individual components—we cannot ignore discrete components completely. Well, actually we can, but you may be curious and want to have a bit more involvement with them. If so, the projects to follow will show you how individual parts are used. The circuits presented here are trivial compared with the sophisticated technology you use every day, such as big-screen LCD HDTV sets, laptops, cell phones, and your wireless Internet broadband router. Feel free to skip these if you wish.

Project 3.1

Setting Up a Simple Lab

If you want to play around with some components just to get the feel for how it used to be, you need to set up a simple lab. If you performed the previous projects, you already have a multimeter. Now you will need a way to interconnect parts in circuits and a power source.

The power source can be four AA cells in a holder. This will provide DC voltage to power the circuits. Or you can use a common 9-volt battery. A battery clip with leads will allow you to connect the battery to the circuit.

Next you will need a breadboard. This is a connector socket with multiple holes in it like that shown in Figure 3.21. You plug the wire leads of the components into the holes to connect them to one another without soldering them.

Then you will need some components such as resistors, capacitors, and transistors. In this project, the goal is to get the battery and the breadboarding socket. The remaining projects for this chapter are circuit experiments.

You can get the breadboard from a local Radio Shack or from a mail order house like Jameco. There are usually two sizes, large and small. The small one is adequate. Figure 3.21 shows a larger one.

Get a battery holder that will take four AA batteries. Again, Radio Shack or Jameco are good sources. Some battery holders also have an on/off switch but it is not necessary. They will have two leads for connecting the battery to the circuit. The 9-volt battery clip is shown in the figure.

Once you get the parts, put the AA cells into the holder so that you connect the negative end (−) of one battery to the positive end (+) of the next so that the cells will be in series. That way their voltages will add together. The circuit should look like that in Figure 3.22.

Next, test the battery. Using your multimeter on the DC ranges, and measure the voltage across the battery holder wires. What do you get? You should read 6 volts or a bit more. The 9-volt battery should, of course, measure 9 volts.

Now you are ready for a few additional component projects.

(Continued)

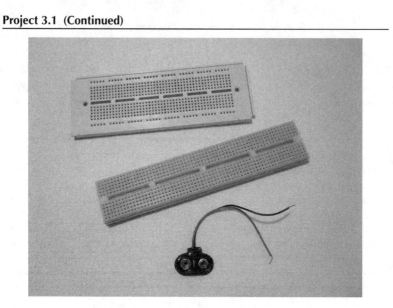

FIGURE 3.21 Common breadboarding socket for building prototypes or just experimenting. *(Courtesy Global Specialties.)*

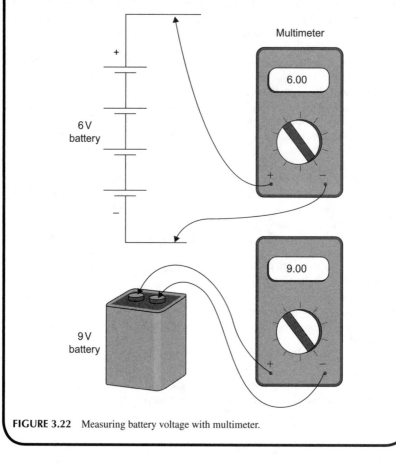

FIGURE 3.22 Measuring battery voltage with multimeter.

Project 3.2

Light Emitting Diode

When you are buying your parts, get an LED and a set of resistors. Any old common red or green LED will work. The resistor set will be a package of all the most common values of resistors.

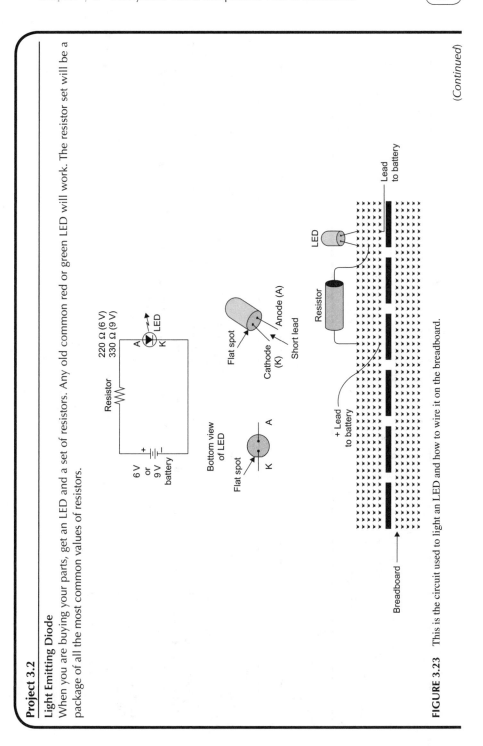

FIGURE 3.23 This is the circuit used to light an LED and how to wire it on the breadboard.

(Continued)

Project 3.2 (Continued)

Now, wire up the circuit in Figure 3.23. A pictorial is shown in the figure as well. Be sure to get the LED leads connected correctly, for if not, you will zap it. See the wiring details in the figure.

The resistor is needed to control or limit the current so that you do not blow out the LED. A value of 220 ohms is a good one if you use the 6-volt battery. The color code for that resistor is red-red-brown-gold. If you are using the 9-volt battery, use a 330-ohm resistor (orange-orange-brown-gold).

Wire the circuit and connect the battery. The LED should light. Now you are an electronic circuit genius.

Project 3.3

Capacitors

Capacitors store a charge and then release it. The charge is stored energy. You charge the capacitor by applying a voltage to it. Then you discharge the capacitor by connecting it to a load like a resistor. This brief project will illustrate that idea.

You will need the following parts from Radio Shack or another source. Get a 100-K ohm resistor (color code brown-black-yellow-gold) and a 100-μF (microfarad) electrolytic capacitor. This capacitor is polarized, meaning that you have to observe the polarity of the voltage applied to it. The wire leads are usually marked with either a + sign, or sometimes with a − sign.

Wire the circuit shown in Figure 3.24 on the breadboarding socket. Do not initially connect the positive lead of the battery to the resistor. Connect your multimeter across the capacitor. Now connect the 6 volts (or 9 volts) from the battery supply as indicated when you are ready to start charging the capacitor. You should see the voltage begin to rise slowly. It should take about 50 seconds for the capacitor to fully charge to 6 (or 9) volts.

Now disconnect the battery's positive lead. The capacitor retains the charge, as you should see on the meter. It will probably begin to drop slowly as it discharges into the meter.

Now use the capacitor as a voltage source. Connect the capacitor to the resistor and LED you used earlier as shown in Figure 3.24. The capacitor will discharge quickly and you should only see the LED blink briefly. To see it again, recharge the capacitor and repeat.

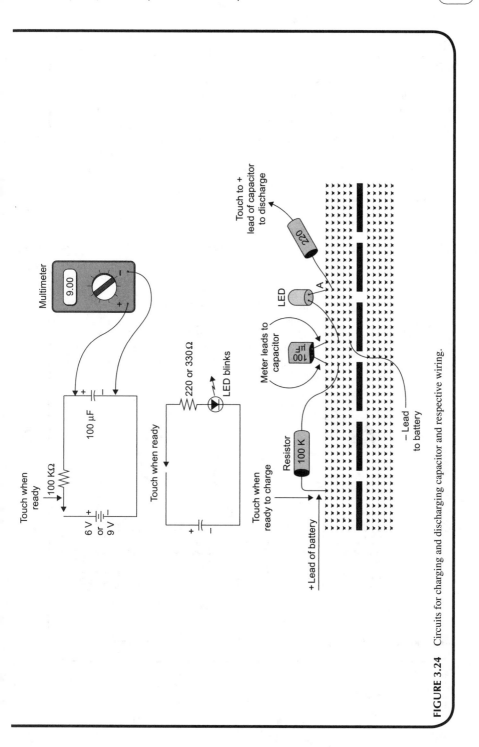

FIGURE 3.24   Circuits for charging and discharging capacitor and respective wiring.

Project 3.4

Transistor Switch

A transistor makes a good switch, especially a MOSFET. And it is easy to use. This project demonstrates how the MOSFET can be used to turn the LED off and on.

Acquire a MOSFET, such as the IRF510 or its equivalent. Then build the circuit shown in Figure 3.25. Be sure to follow the guidelines for wiring the transistor. Use the LED and 220- (or 330-) ohm resistor that you used before.

Now, connect the battery. Touch the wire from the gate on the MOSFET to the +6 volt or +9 connection like on the resistor. This will turn the transistor on and it will act as a switch to turn on the LED. Disconnect the gate from +6 volts and the transistor switch goes off, turning off the LED.

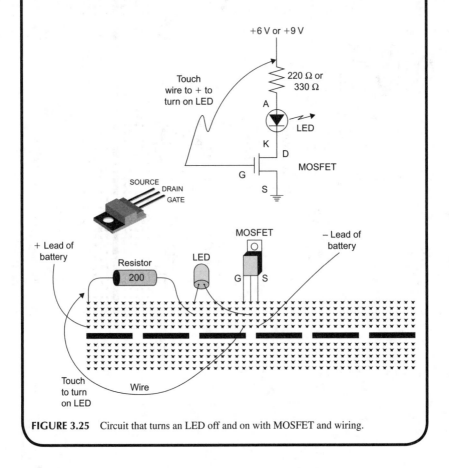

FIGURE 3.25 Circuit that turns an LED off and on with MOSFET and wiring.

Chapter 4

Electronic Circuits: Linear/Analog

The Building Blocks of Electronic Equipment

In this Chapter:
- Linear circuits defined.
- Types of linear circuits.
- Characteristics and specifications of the most common linear circuits.

INTRODUCTION

The basic building blocks of electronic equipment are individual circuits made up of transistors, resistors, capacitors, and sometimes transformers, inductors or other components. It used to be that circuits were made up of individual discrete components that were wired together on a printed circuit board. But today, most electronic circuits are in integrated circuit (IC) form or more likely multiple circuits are already combined on a single silicon chip to create a larger circuit, sub-system or system on a chip (SoC).

While most circuits are ICs there are still some cases where discrete components are necessary or desirable. The most common examples are high voltage or high power circuits that would burn up a tiny silicon chip. Discrete components have not gone away entirely and you will find some of them external to an IC such as larger capacitors or inductors that just cannot be easily integrated on-to a chip or transistors that can handle higher voltage or power. As mentioned before, we are not going to talk about the make-up or operation of individual circuits, Instead, this book treats circuits as just building blocks with an input and an output that process the input in some way to create a new, different output.

LINEAR CIRCUITS

There are two basic types of circuits, linear and digital. Digital circuits are covered in the next chapter. Linear circuits are those circuits that process analog signals. Recall that analog signals are those that vary smoothly and continuously

doi: 10.1016/B978-1-85617-700-9.00004-7

over time in contrast to pulse and digital signals that are voltages that switch off and on very fast. Analog signals are like sine waves, voice signals, video signals, radio waves, and voltages from sensors like a temperature transducer. DC, either a constant value of voltage or one that is varying, is also an analog signal.

A linear circuit is one whose output is proportional to its input. See Figure 4.1. Its so-called response is a straight line or what a mathematician would call a linear function. The best example of this is the amplifier where its output is proportional to its input so that it accurately reproduces the input faithfully but at a higher voltage or power level. In the figure, with 0.4 volt input you get 4 volt output. That means the amplifier has a gain of 10. Linear circuits do not distort the signal. This chapter covers linear circuits with most of the emphasis on amplifiers since that is the largest category of linear circuits. And as it turns out, most other types of linear circuits are just amplifiers connected in different ways to give different processing results.

Amplifiers

The basic block diagram symbol for an amplifier is a triangle. See Figure 4.2. It has an input and an output. Amplifiers also have one or more DC inputs as well. This DC from a battery or other power supply is what powers the circuit

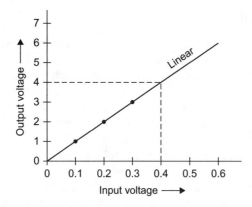

FIGURE 4.1 Linear circuit has a straight-line response. As input varies, output varies by factor of 10 larger, producing straight-line output.

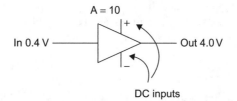

FIGURE 4.2 Block symbol for amplifier. Gain, A, is often given. DC power inputs are not usually shown.

and it is this DC that is ultimately converted into the new, larger output signal. In most cases the DC inputs are not shown in a block diagram. The amplifier processes the input to create an output of the same shape but at a larger amplitude.

The primary characteristic of any amplifier is its *gain*. The gain (usually represented by A) is simply a number equal to the ratio of the output to the input.

$$A = \frac{\text{Output}}{\text{Input}}$$

If the output is 4 volts and the input is 0.4 volt, the gain is simply:

$$A = \frac{4}{0.4} = 10$$

The amplifier multiplies the input by 10 to get the output.

$$\text{Output} = 10 \times \text{Input}$$

Classifying Amplifiers

The amplifier is the most common linear circuit. There are so many different types of amplifiers that we divide them up into different types to distinguish one from another. And there are multiple ways of doing this. Here are the most common.

Small Signal versus Large Signal

Small signal amplifiers are those that, as the name implies, only amplify small signals which are those roughly below about 1 volt. Small signal amplifiers amplify millivolt, microvolt, or nanovolt signals. Large signal amplifiers amplify larger signals and typically amplify the power of a signal rather than the voltage. These are called power amplifiers (PAs).

Frequency Classification

Amplifiers are also categorized by the frequency of the signals they amplify. Some examples are DC amplifiers, audio frequency (AF) amplifiers, intermediate frequency (IF) amplifiers, video amplifiers, radio frequency (RF) amplifiers, and microwave amplifiers.

Configuration Classification

The two main configurations are single-ended and differential. A single-ended amplifier is one in which the input and output are referenced to a common ground. These amplifiers have two inputs, one is the signal and the other is ground. See Figure 4.3A. Normally only the input is shown as the ground is just assumed to be there.

A differential amplifier has two inputs as Figure 4.3B shows. These are usually called the inverting (−) and the non-inverting (+) inputs. Both inputs are referenced to ground. The amplifier is called a differential amplifier because it actually takes the mathematical difference between the two inputs and amplifies the result. One input is subtracted from another, then the difference is amplified by the gain A. The formula for this operation is:

$$\text{Output} = (\text{Input } 2 - \text{Input } 1) \times A$$

Differential amplifiers have one neat feature in that any signal common to both inputs like noise is automatically canceled out. This feature is widely taken advantage of in very small signal amplifiers used in memory chips and high gain amplifiers that amplify small signals from sensors. Noise that is picked up simultaneously on both inputs simply subtracts itself out without bothering the desired small difference signal.

Classification by Class of Operation

One way to categorize amplifiers is to designate the conduction of the transistors in the amplifiers. The primary classes are A, B, AB, C, D, E, and F. There are a few others but these are the most common. Class A amplifiers conduct continuously. We say that the amplification occurs over the full 360 degrees of a sine wave cycle. See Figure 4.4. Note: The small t on the waveforms means time as the voltage is varying over time.

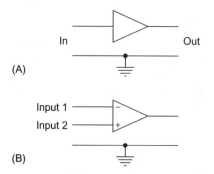

FIGURE 4.3   Single-ended (**A**) and differential input (**B**) amplifiers. Inputs and outputs are referenced to ground.

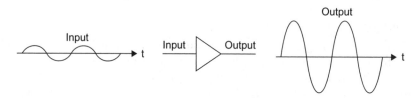

FIGURE 4.4   Class A amplifier conducts continuously.

Class B amplifier transistors only conduct for half a cycle or 180 degrees of a sine wave. See Figure 4.5. Such an amplifier seems useless since it distorts the signal. But if you combine two transistors, one to amplify positive cycles and the other to amplify negative cycles, then combine the two, the complete wave is amplified. Such an amplifier is called a push pull amplifier. The benefit of this arrangement is that the circuit is more efficient.

Class AB amplifiers are a variation of the class B. Class B amps do add a little distortion to the signal around the point where the sine wave crosses zero. This is caused by the transistors not turning on and off precisely at the zero point. This glitch shown in Figure 4.5 is called crossover distortion. To get rid of that, class AB amplifiers allow some small current to flow continuously. The result is less distortion but a little less efficiency.

Class C amplifiers are for RF signals and are usually power amplifiers. They only allow the transistor to conduct for less than 180 degrees of a sine wave input, usually 90 to 150 degrees. It produces massive distortion. However, the amplifier output is made up of an inductor and capacitor (LC) tuned circuit that oscillates at the operating frequency and removes the distortion. These amplifiers are very efficient.

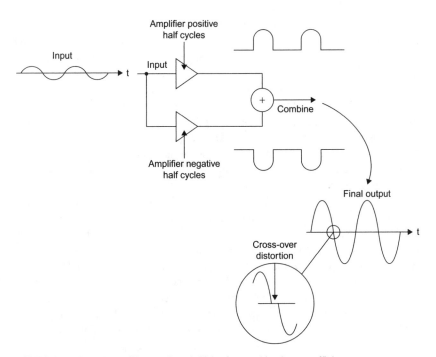

FIGURE 4.5   Class B amplifier conducts half the time, making it more efficient.

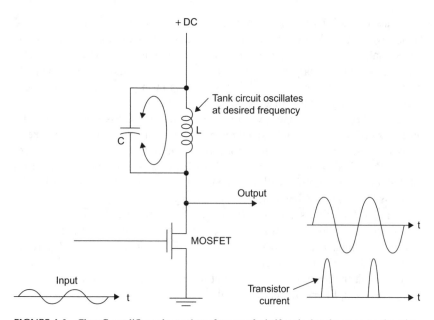

FIGURE 4.6   Class C amplifier only conducts for part of a half cycle, but the current pulse stimulates resonant tank circuit that oscillates at desired frequency.

In a class C amplifier the MOSFET acts like a switch that is turned off and on by the input signal, usually a sine wave. See Figure 4.6. When the MOSFET is on the capacitor charges up to the DC supply voltage and current flows in the inductor creating a surrounding magnetic field. When the transistor turns off, the inductor and capacitor begin to exchange energy and set up an oscillation at the resonant frequency of the LC circuit. This is called a tank circuit. The result is that the energy stored in the tank circuit produces the amplified sine wave output.

Another way of looking at this is that the transistor switch distorts the input creating a pulse waveform that contains lots of harmonics. The tank circuit acts like a selective band-pass filter that lets the fundamental sine wave pass while the harmonics are filtered out.

Don't forget the Fourier theory and harmonics. Remember that a harmonic is an integer multiple of a fundamental sine wave. If the fundamental sine wave is 20 MHz, the 2nd harmonic is 40 MHz, the 3rd harmonic is 60 MHz, the 4th harmonic is 80 MHz, and so on. Any non-sine wave signal, like a half sine wave, square wave, or pulse is said to be made up of the fundamental sine wave added to multiple harmonics. A distorted sine wave has the fundamental sine wave in it but also harmonics. This concept is called the Fourier theory after the French mathematician.

Class D amplifiers are a special class called a switching amplifier. It is made up of transistor switches rather than a real linear amplifier. This amplifier chops the input analog signal into high-frequency pulses with a varying width. See Figure 4.7. This process is called pulse width modulation (PWM).

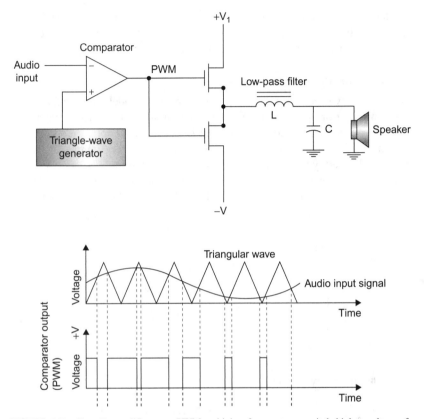

FIGURE 4.7   Class D amplifier uses PWM at higher frequency to switch higher voltages for greater output. Class D amplifiers are more efficient than any other type, and thus do not produce as much heat.

The sine wave audio signal to be amplified is sent to one input of a special amplifier called a comparator along with a higher-frequency triangle wave. The comparator output switches when the triangle wave and sine wave values are equal. The resulting PWM signal is then fed to MOSFET switches to make the signal bigger. The higher amplitude output it then filtered back into an analog signal by a low-pass filter made up of a capacitor and inductor. Most class D amplifiers are audio power amplifiers whose load is a speaker or headphone. For low power less than a few watts, all the circuitry is in a single IC. For higher power, the MOSFETs are larger external devices.

The big benefit of switching amplifiers is that for a given output power they are very efficient. Whereas a class AB amplifier may only be 20 to 30% efficient, the class D is over 90% efficient. This means that the amplifier can be smaller and use less power and dissipates less heat. Class D amplifiers are great for portable battery-operated devices like MP3 players and cell phones.

Classes E and F are special switching amplifier variations used at RF frequencies only. Like the class C amplifier, they use LC circuits to filter out the harmonics they produce. They are very efficient.

Operational Amplifiers

There are many special types of amplifiers. The most widely used is the operational amplifier or op amp. This is a versatile high-gain DC differential amplifier that can be configured with external resistors, capacitors, diodes, etc. to perform a wide variety of amplification functions. The op amp may be the most widely used type of amplifier because of its versatility.

Figure 4.8 shows the most common op amp configurations. In most cases, the resistors are external to the IC amplifier. Note that the gain is determined by the values of the external resistors. Note formulas for gain (A).

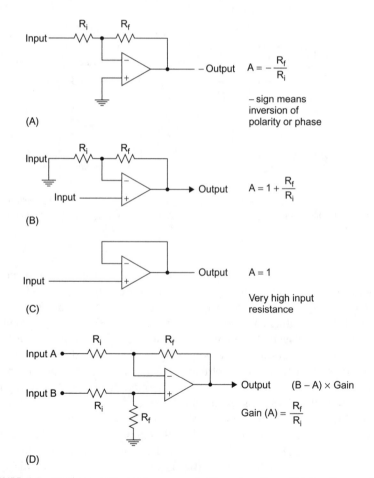

FIGURE 4.8 Most commonly used op amp circuit configurations. **(A)** Inverting amplifier. **(B)** Noninverting amplifier. **(C)** Follower. **(D)** Differential amplifier.

A variation of the op amp is the instrumentation amp. See Figure 4.9. It is a combination of multiple op amps that make it very useful in industrial applications such as amplifying small signals from sensors. Again the gain is determined by the resistor values. You can make this amplifier from individual op amps but usually you just buy it as a single integrated circuit. The external resistor sets the gain.

A variable or programmable gain amplifier is one whose gain can be controlled from an external DC input.

Amplifier Specifications

While you really do not need to know much about what the circuitry in an amplifier IC is, you do need to understand the specifications. Here are the most common specifications to look for.

As mentioned before, gain is the primary amplifier characteristic. It is usually a voltage or power ratio, depending on the type of amplifier.

$$\text{Voltage gain is output voltage over input voltage: } Av = \frac{Vo}{Vi}$$

$$\text{Power gain is output power over input power: } Ap = \frac{Po}{Pi}$$

Many amplifiers have their gain expressed in decibels (dB). A dB is a gain measurement based on a logarithmic scale rather than a linear scale. This has the advantage of being able to express large and small gains in smaller numbers.

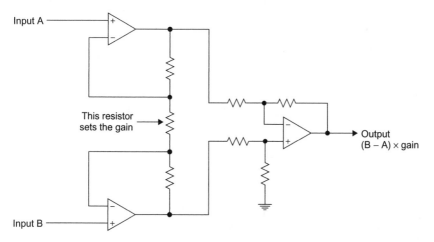

FIGURE 4.9 Instrumentation amplifier widely used to amplify small sensor signals in industrial applications.

And gains of multiple stages can be added rather than multiplied. To calculate dB gains you used these formulas:

$$\text{Voltage gain: } dB = 20 \log (Av)$$

$$\text{Power gain: } dB = 10 \log (Ap)$$

In these formulas, log means the common (base 10) logarithm of the voltage or power gain. The logarithm (log) is easy to find using a scientific calculator.

For example, a voltage gain of 50 expressed in dB is:

$$dB = 20 \log 50 = 20 (1.7) = 34 \text{ dB}$$

A power gain 300 expressed in dB is:

$$dB = 10 \log 300 = 10(2.5) = 25 \text{ dB}$$

Input Impedance

Input impedance is the load seen by the signal source to the amplifier. It is usually resistive but typically also has some capacitance across it. Figure 4.10A shows an equivalent circuit of an amplifier. Note the input resistance. That is the input impedance. It may be as high as several megohms or could be as low

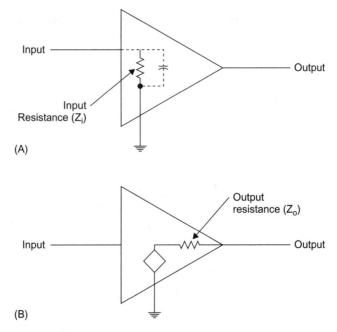

FIGURE 4.10 Amplifier input impedance (**A**) and output impedance (**B**).

as, say, 50 ohms. Usually the higher the better for voltage amplifiers but not for power amplifiers.

Impedance

The term *impedance* means opposition to current flow. It is actually the combined opposition offered by resistance plus any capacitive or inductive reactance. If there is no reactance, the impedance is just the resistance so the two terms can be used interchangeably. Impedance is represented by the letter Z.

Output Impedance

Amplifiers act as signal generating sources because they essentially just create a new higher voltage or higher power signal with the same shape as the input. They appear as a signal generator to the next circuit or load. Figure 4.10B shows the amplifier as a generator shaped as a diamond. It has an internal impedance or series resistance called the output impedance. It is not a real resistor but it is there and it appears in series with whatever load is connected which may be another amplifier or something else.

Figure 4.11 shows the amplifier connected to another amplifier. Note that the output impedance for amplifier A_2 and the load form a voltage divider so some of the amplification is lost to the voltage divider effect. The input impedance of A_2 is the load for the output impedance of A_1.

Cascading Amplifiers

If an amplifier does not have enough gain, then multiple amplifiers can be cascaded one after the other to give the needed gain. One amplifier amplifies the output of the next and so on. See Figure 4.12. To calculate the overall gain of an amplifier you just multiply the gains as shown in the figure. This procedure assumes that one amplifier does not load the next. That is, the output impedance of each amplifier is assumed to be zero and the input impedance of the next amplifier is assumed to be infinite. That way, there is no voltage divider effect that lowers the gain.

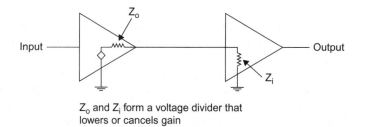

Z₀ and Zᵢ form a voltage divider that lowers or cancels gain

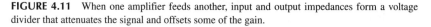

FIGURE 4.11 When one amplifier feeds another, input and output impedances form a voltage divider that attenuates the signal and offsets some of the gain.

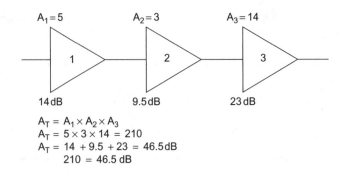

$$A_T = A_1 \times A_2 \times A_3$$
$$A_T = 5 \times 3 \times 14 = 210$$
$$A_T = 14 + 9.5 + 23 = 46.5\,dB$$
$$210 = 46.5\,dB$$

FIGURE 4.12 Cascading amplifier provides more gain. Total gain is product of individual gains or sum of dB gains.

If the amplifier gains are given in dB, you just add the dB figures to get the total gain.

Efficiency

Efficiency is the ratio of the output power to the input power expressed as a percentage.

$$\% \text{ Eff} = \frac{Po}{Pi} \times 100$$

An amplifier takes the DC supply voltage and converts it into the AC output. The efficiency expresses how well the amplifier does this. Any lost power usually gets converted to heat which is not good. Nevertheless, some heat will always be generated just because electronic circuits are not perfect. High efficiency is a good thing of course, but often efficiency is traded off for better linearity, fidelity, and lower distortion of the signal.

Class A amplifiers have the poorest efficiency commonly in the 10 to 30% range. Classes B and AB are more efficient in the range of 50 to 70%. Class C amps have an efficiency of 70 to 85%. Class D amplifiers are the best with efficiencies in the 80 to 95% range.

Frequency Response

This is a measure of the range of frequencies that amplifiers cover. DC amplifiers may only cover DC up to about a few hundred Hz. Audio amplifiers may cover voice frequencies from 300 to 3000 Hz or stereo amps that cover 20 Hz to 20 kHz. IF amplifiers may only cover a specific band, say, 455 kHz, 9 MHz, 45 MHz, 70 MHz, or some other popular IF value. Op amps amplify signals from DC up to some very high frequency as high as several hundred MHz in some cases.

RF amplifiers cover high radio frequencies from a few MHz up to microwaves from 1 to 100 GHz. Their frequency and range of coverage is usually set by a filter or LC tuned circuit.

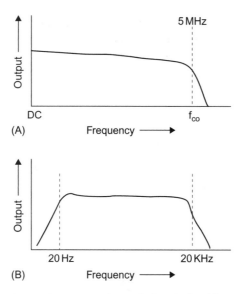

FIGURE 4.13   Frequency-response curves show output versus input frequency. Curve in (**A**) is response of op amp that extends from DC to 5-MHz cut-off frequency. Curve in (**B**) is response of common stereo audio amplifier.

Figure 4.13A shows the response curve for an op amp with an output down to DC and up to the cut-off frequency (f_{co}) or 5 MHz. Figure 4.13B is the response curve of a stereo amplifier with a range of 20 Hz to 20 kHz.

Output Capability

This specification indicates the range of output signals the amplifier is capable of. It is usually stated as a maximum voltage swing or as a maximum amount of output power. In both cases, the maximum output is usually given for a given value of load resistance. For example, an op amp may be capable of producing an output of ± 12 volts across a 2 K load. A power amplifier may produce a maximum output of 10 watts across an 8 ohm speaker load or, say, 50 watts across a 50 ohm antenna load.

Filters

A filter is a frequency sensitive circuit that is designed to pass some frequencies, but reject others. There are various ways of making filter circuits that will accept some frequencies, but greatly attenuate others. There are literally thousands of applications for such circuits. But the main use of filters is to retain a desired signal, but eliminate interference and noise.

There are four basic types of filters: low-pass (LPF), high-pass (HPF), band-pass (BPF), and band-reject (BRF). The names, of course, tell their function.

A low-pass filter (LPF) is a circuit that passes all frequencies between DC and some upper cut-off frequency (f_{co}). All frequencies above the cut-off frequency are rejected. Figure 4.14A shows the response curve of an LPF. This is a plot of the filter output amplitude versus frequency. Note the gradual slope

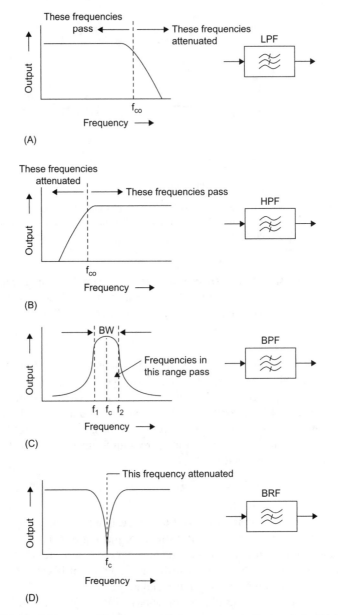

FIGURE 4.14 Four common filter types and their response curves and symbols. (**A**) Low-pass. (**B**) High-pass. (**C**) Band-pass. (**D**) Band-reject or notch.

as signals above the cut-off frequency (f_c) are reduced in amplitude. The block symbol is how LPF is represented in some diagrams.

A high-pass filter passes frequencies above the cut-off and greatly attenuates those below the cut-off. Figure 4.14B shows the ideal response curve of an HPF. Note the HPF symbol.

Band-pass filters (BPF) pass a certain range of frequencies, but reject all those above and below that range. Figure 4.14C shows the practical response curve of a BPF. Note that there are two cut-off frequencies, one above and one below the center frequency f_c. The difference between the upper and lower cut-offs is called the filter bandwidth (BW):

$$BW = f_2 - f_1$$

This is the range of frequencies passed. Again note the schematic symbol for a BPF.

Figure 4.14D illustrates the curve for a band-reject or notch filter. It is used to eliminate one single frequency and those close around it.

Note in the block symbols in Figure 4.14 that each has three waves representing high, middle, and low frequencies. The wave with the slash through it means that these frequencies are cut out or greatly attenuated.

Filters can be implemented in many ways. Resistor and capacitor (RC) filters are common as are inductor and capacitor (LC) filters. You can make filters with op amps called active filters. Other filters are made with mechanical vibrating elements, a ceramic or quartz crystal, or a special surface acoustic wave (SAW) filter. Many filters today are made with digital signal processing (DSP). These will be discussed in Chapter 6.

Oscillators

An oscillator is a circuit that generates a signal. It is usually a sine wave or a rectangular wave. The electronic symbols used to represent an oscillator are given in Figure 4.15.

The main specification of an oscillator is its output frequency. Most oscillators have a fixed frequency but there are oscillators whose frequency can be varied. A tunable frequency oscillator is called a variable frequency oscillator (VFO). You can often tune it by turning a knob that varies a capacitor or inductor to change the frequency. Some oscillators are tuned by an external voltage. These are called voltage-controlled oscillators (VCOs). By varying the value of a DC input voltage, the frequency is changed.

The frequency of an oscillator is set by some frequency determining element or circuit. That circuit may be made up of resistors and capacitors (RC), inductors and capacitors (LC), or a crystal. A crystal is a thin piece of quartz that can be made to vibrate at a very precise frequency. It maintains its frequency very closely unlike RC or LC networks that can drift with temperature or change from vibration or other conditions. Crystals come in a small metal package and are used to set the frequency in most oscillators today.

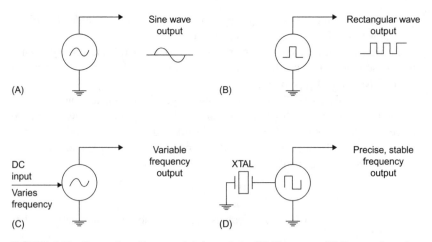

FIGURE 4.15 Types of oscillators and their symbols. **(A)** Sine wave. **(B)** Rectangular pulse. **(C)** VCO. **(D)** Crystal clock oscillator.

A special type of oscillator is called a clock oscillator. It is used as an accurate time and frequency source. The frequency determining element is usually a quartz crystal that vibrates at a precise frequency and is very stable. The clock signal is used to operate all the circuits in synchronism. An example is the clock oscillator in a PC that sets the speed of calculations. Sometimes you will see a quartz crystal (XTAL) symbol drawn with the oscillator symbol to illustrate this. See Figure 4.14D.

Mixers

A mixer is a circuit that takes two input signals and mixes them in a special way to produce new output signals. There are two types of mixers, linear and non-linear.

A linear mixer is one that takes the input signals and just adds them together. You can do this with some resistors as shown in Figure 4.16A. The sum of both input signals appears across the output resistor.

You can also make a linear mixer with an op amp as shown in Figure 4.16B. This op amp circuit is called a summer. Setting the resistor values allows you to mix and add gain at the same time. A good example of a linear mixer is those used by musicians so they can mix the outputs from microphones and musical instruments so they can all be amplified as one by the power amplifiers that drive the speakers. Note the variable resistors called pots that are used as amplitude or volume controls.

A non-linear mixer takes two input signals and effectively multiplies them together. The result is that the mixer produces four output signals. Figure 4.17 shows the symbol for a mixer. The inputs are frequencies 1 and 2 or f_1 and f_2.

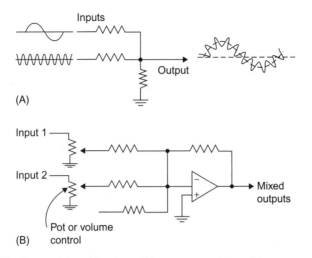

FIGURE 4.16 Linear mixing with resistors (**A**) or an op amp summer (**B**).

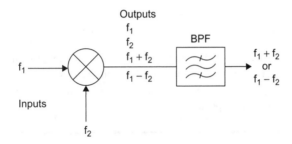

FIGURE 4.17 Non-linear mixer used to generate higher and low frequencies.

The mixer outputs are f_1, f_2, and ($f_1 + f_2$) or the sum, and ($f_1 - f_2$) the difference. The sum and difference frequencies are the ones of most interest. Usually a filter is used at the output of the mixer to select the desired frequency and eliminate the others. If the lower difference frequency is selected, the mixer is often called a down converter. If the sum frequency is selected, the mixer is called an up converter. The mixer is a way to translate a frequency to another higher or lower frequency that is better for processing. Any signal amplitude or frequency variations on the inputs are accurately carried through to the new output frequency.

Phase Detector

A phase detector is a mixer-like circuit that puts out a signal that is proportional to the phase difference between two input signals of the same frequency. See Figure 4.18. A phase shift is a time difference between two signals of the

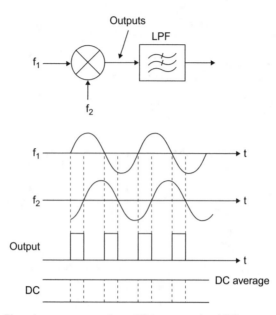

FIGURE 4.18   Phase detector converts phase shift into proportional DC average.

same frequency. We sometimes need to know how much phase shift is present. The phase detector produces a series of output pulses whose width is proportional to the phase difference. Passing the pulses through a low-pass filter smoothes them into a proportional DC voltage.

Filtering Pulses into DC

A common occurrence in electronic circuits is the filtering of pulses into a DC voltage. You saw this back in Figure 3.15 where positive pulses from a sine wave rectifier are smoothed into a DC voltage. It is the capacitor that charges up and stores the pulse voltage. Then between pulses the capacitor discharges into the load maintaining a near constant DC voltage. The low-pass filter usually includes a capacitor so performs the same function in the phase detector of Figure 4.18. The output is a proportional average DC output.

Phase-Locked Loops

The phase-locked loop (PLL) is a frequency- and phase-sensitive feedback control circuit. It consists of three major parts: a mixer or phase detector, a low-pass filter (LPF), and a voltage-controlled oscillator (VCO), as shown in Figure 4.19. The signal from the VCO is compared to the input signal. If there is a frequency (or phase) difference between the two, an error signal is generated. This error signal is filtered by the LPF into a varying DC level and is used to control the VCO frequency. This is the feedback signal.

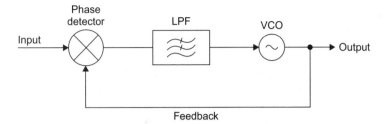

FIGURE 4.19   Phase-locked loop.

If the VCO frequency, f_o, differs from the input reference signal frequency, f_o the f_r the phase detector sees this as a phase shift. This causes the phase detector to produce an error signal. This error signal is filtered into a varying DC signal that is used to control the VCO.

The VCO can be either a sine wave oscillator or a rectangular wave oscillator depending upon the desired waveshape of the output. In either case, the VCO output frequency is made directly proportional to the DC control voltage. There is a linear voltage-to-frequency relationship in the VCO. This means that changing the control voltage produces a proportional change in output frequency. If the voltage goes up, so does the frequency.

Now if an input reference signal whose frequency is near that of the VCO is applied to the PLL, the phase detector will produce an output voltage proportional to the frequency difference. This signal is filtered and the resulting DC control voltage is applied to the VCO. The control voltage is such that it forces the VCO frequency to move in a direction that reduces the error signal. This means that the VCO frequency will change until it is equal to the input reference signal frequency. When this happens, the two signals are synchronized or "locked." The phase difference causes the phase detector to produce the DC voltage at the VCO input to keep the PLL locked to the input signal.

If the input reference signal changes, then the phase detector will recognize a frequency (or phase) difference between the input and the VCO output. As a result, the LPF will produce a different DC control signal that will force the VCO to change such that it is equal to the new input frequency. As you can see then, the PLL will "track" an input signal frequency as it changes.

The range of frequencies over which the PLL will track an input signal and remain locked is known as the lock range. This is a range of frequencies above and below the VCO free running frequency. The PLL can track and "lock" to any input frequency in this range. If an input signal out of the lock range is applied, the PLL will not synchronize.

If the input signal is initially outside of the lock range, the PLL will not lock. But, the PLL will jump into a locked condition as soon as the input frequency gets close to the VCO frequency. In other words, the PLL will "capture" the incoming signal if it is close enough to the VCO frequency. Once the input signal is captured, the PLL is locked and will track further changes in the input signal frequency.

The range of frequencies over which a PLL can capture a signal is known as the capture range. Like the lock range, it too is centered on the free running frequency. But the capture range is narrower than the lock range. The PLL acts as a frequency sensitive circuit over a narrow range of frequencies.

Since the PLL will only capture and lock on to input signals within a certain narrow band, the PLL acts like a band-pass filter. For that reason, the PLL is an excellent signal conditioner. You can take a noisy input signal or one with undesirable interference on it and filter it with a PLL. The PLL will lock on to only the desired frequency component of the input. The VCO reproduces the input signal at the same frequency but with the noise and interference removed. The PLL not only cleans up a signal but also can track it if its frequency changes.

The PLL is widely used for a variety of purposes. It is used to recover the clock signal in some wireless applications. It is used to recover the original signal in frequency modulation (FM) radio. It is used to multiply a frequency by a fixed factor. It can be used for motor speed control. Almost all electronic products of some kind contain a PLL.

Frequency Synthesizers

One of the most popular uses of the PLL is in frequency synthesizers. A frequency synthesizer is a very stable signal source that can be varied over a specific range of frequencies in finite increments or steps. Frequency synthesizers are used as precision signal generators for test and measurement purposes and replace more conventional oscillator circuits as the primary signal sources in radio transmitters or receivers.

The big advantage of a PLL synthesizer is its ability to generate a wide range of frequencies with great accuracy and stability by using only a single stable signal source. For example, a PLL synthesizer with a single crystal controlled reference input can readily generate hundreds or even thousands of discrete frequencies that are as stable as the crystal reference. By using a synthesizer as the initial signal source in a transmitter, a wide range of channels can be obtained with a single crystal. When many channels must be used, a synthesizer greatly reduces the cost since crystals are expensive.

The synthesizer circuit in Figure 4.20 shows that it is a PLL with a frequency divider connected between the VCO output and the phase detector. The frequency divider is a circuit that divides its input frequency by some whole number. For example, the output of a divider with a division factor of 12 and an input of 48 MHz is 48/12 = 4 MHz.

The frequency synthesizer PLL also uses a stable, crystal oscillator input to the phase detector. This is usually derived from a crystal oscillator. When the PLL is locked, the stability and accuracy of the VCO output will be the same

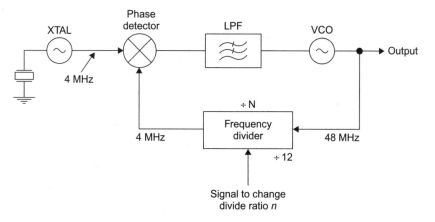

FIGURE 4.20 Frequency synthesizer using phase-locked loop. Changing the divider ratio changes the frequency.

as that of the reference input. The frequency divider is usually made so that its divide ratio can be changed by setting a switch or changing the control inputs. Changing the division ratio changes the output frequency at which the loop will lock.

The reason for this is that the two inputs to the phase detector must always be equal in frequency if a locked condition is to be achieved. The reference frequency is fixed so the only thing that can change is the VCO output if the divider ratio is altered. If the loop is locked, the output of the divider will be 24 MHz no matter what the divide ratio is. Therefore, with an output of 4 MHz, a divide by 6 circuit must have a 24 MHz input. Since the VCO output is the divider input, the VCO and synthesizer output will be 24 MHz too.

As you can see, changing the divide ratio changes the output frequency. In this circuit, if the divide ratio is changed in integer values between 6 and 12, the output frequencies will be as shown in Table 4.1. With this setup, the output is switched in 4-MHz increments.

If the input reference is made 1 MHz, then when the divide ratio is changed, the output frequency will step in 1-MHz increments. By carefully selecting the reference frequency and the divide ratio, it is possible to synthesize practically any range of frequencies in many increments from subaudio to microwave.

Note one final fact about the PLL circuit in Figure 4.20. If the reference is considered the input, and the VCO is considered the output, then the PLL is acting as a frequency multiplier. Here the output is 12 times the input. You will occasionally see the PLL used as a frequency multiplier in some applications.

The PLL synthesizer is a very versatile circuit. It is commonly available in integrated circuit form. PLL frequency synthesizers are used in TV channel selectors, stereo receivers, CB radios, cellular telephones, and other two-way radios. In such equipment the divider ratio is supplied by an input keyboard or

TABLE 4.1 PLL Divide Ratio and VCO Output

Divide Ratio	VCO Output
6	24 MHz
7	28 MHz
8	32 MHz
9	36 MHz
10	40 MHz
11	44 MHz
12	48 MHz

a built-in microcomputer programmed to supply the desired divide factor for a selected output frequency.

Power Supplies

A power supply is, as its name implies, a source of power to an electronic circuit. Most electronic circuits operate from DC as they process AC or DC signals. The most common power supply is the one that gets its input from the standard AC wall outlet that supplies 120 volts at 60 Hz. This common voltage is then converted by the power supply into one or more DC voltages that go to operate the TV set, computer, or other equipment. This section discusses the main components and circuits of a power supply.

Batteries

One of the most widely used power supplies is a battery. A battery is a great source of DC by itself and no AC source is needed. Batteries were the very first form of voltage sources for electrical circuits and they are still widely used today. What would we do without the batteries for our cell phones, iPods, cordless telephones, laptop computers, and remote controls? Just remember that a special form of power supply is the battery charger that normally uses AC to recharge lead-acid, nickel cadmium, nickel metal hydride, and lithium-ion batteries.

Standard Power Supply

A very common AC to DC power supply is shown in Figure 4.21. It normally consists of a transformer that translates the 120 volt AC line voltage into a higher or lower voltage as needed by its step up or step down characteristics. Today, since most equipment uses ICs, a lower AC voltage is needed. The lower

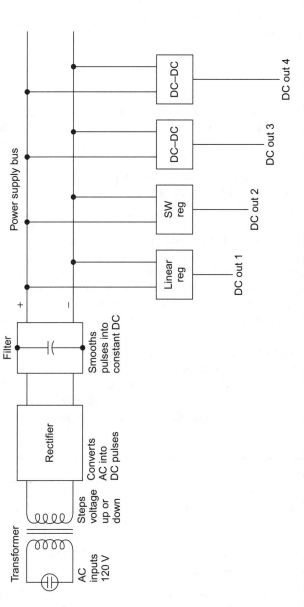

FIGURE 4.21   Typical power supply configuration using bus architecture, regulators, and DC–DC converters to get desired number of outputs. reg, regulator.

AC voltage is then converted into a pulsating DC voltage by a rectifier. A rectifier is one or more diodes that act like polarity sensitive switches to change the positive and/or negative sine half cycles into pulses of DC. A filter is then used to smooth the pulses into a more constant DC voltage. The filter is usually a large capacitor that charges up to the peak of the DC pulses then only discharges very slowly into the load.

Regulators

A regulator is a circuit that maintains a fixed DC output voltage of a desired value. The DC produced by the power supply at the filter output will vary over a wide range as the AC input line voltage changes. Any changes in the load meaning variations in current taken from the supply will also cause the output voltage to vary. A regulator absorbs those variations and creates a constant DC output suitable for powering ICs and other electronic circuits. Common DC output voltages are 3.3, 5, 12, 15, and 24 volts. Other typical DC supply voltages are 1.2, 1.8, 6, and 48 volts.

There are two types of regulators, linear and switching. A linear regulator inserts a transistor in series between the power supply output and the load. If the load varies, circuitry in the regulator adjusts the conduction of the transistor so that a constant voltage is maintained. It is as if a resistor in series with the load forms an adjustable voltage divider. If the output voltage drops, the series transistor conducts more to keep the output voltage the same. If the output increases, the transistor conducts less to decrease the voltage and maintain the same value.

The other type of regulator is the switching regulator. It puts a transistor in series with the load as well but the internal circuitry turns the transistor off and on at a high rate of speed. By varying the duty cycle (ratio of on to time for one period) of the on–off pulses the average DC output may be varied. Then if any output voltage changes are sensed, the duty cycle is adjusted to ensure that the average output stays the same.

Duty Cycle and PWM

The duty cycle is the ratio of the pulse width to the period of a rectangular wave as shown in Figure 4.22. The period is the reciprocal of the frequency of the signal or $T = 1/f$. It is constant. The on-time is the duration of the pulse voltage (t). The pulse duration can vary depending upon what the application is. If the period is 50 milliseconds and the pulse on-time is 10 milliseconds, the duty cycle (D.C.) is:

D.C. = 10/50 = 0.2 × 100 = 20%

Circuits can be made to change the on-time so that the duty cycle can be varied from a few percent to nearly 100%. This is called pulse width modulation (PWM). It was introduced earlier in the class D amplifier discussion. PWM is widely used in switching power supplies like regulators and DC–DC converters.

(Continued)

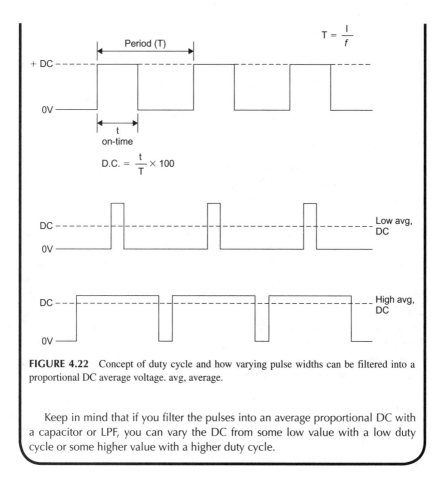

FIGURE 4.22 Concept of duty cycle and how varying pulse widths can be filtered into a proportional DC average voltage. avg, average.

Keep in mind that if you filter the pulses into an average proportional DC with a capacitor or LPF, you can vary the DC from some low value with a low duty cycle or some higher value with a higher duty cycle.

Linear regulators work very well but are not very efficient as they waste power in the series transistor. Typical efficiency is 10 to 40%. A switching regulator is far more efficient as it is off for a large part of the time. Efficiencies in the 70 to 95% range are common.

DC—DC Converters

A DC–DC converter is a circuit that does what its name implies. It translates one DC voltage to another. It can take the voltage from a 12 volt battery and generate several hundred or thousand volts. Or it can take a 4.2 volt battery and produce voltages of 3.3 or 2.5 volts and 15 volts. By building one DC power supply, several other output voltages can be obtained by using different DC–DC converters.

Bus-Oriented Architecture

Many power supplies have what we call a bus architecture. See Figure 4.21. The DC power supply produces one main output voltage. That voltage is applied to a common DC bus. A bus is just the name for the two + and − DC lines that carry the voltage. Then, regulators and DC–DC converters are used to generate the voltages needed by the equipment. Again see Figure 4.21.

Power Management

Many power supplies today include a circuit that manages the power supply voltages. It is usually a single IC that does things like turn on the various DC voltages in a power supply in a specific sequence. Some ICs require multiple DC voltages and they must be turned on one at a time in a specific sequence. In addition, the power management chip does things like monitor current usage. If one power supply is not being used, meaning current not being drawn from it, that part of the power supply is shut down to conserve power. A good example of this is the power management in a laptop computer or cell phone that monitors and controls all DC power to preserve the battery charge for longer life. Some other functions performed by power management chips are voltage monitoring with voltage alarm if a voltage fails, temperature monitoring, fault detection, over voltage or under voltage detection, and power control. Some power management chips also have the battery charging circuits built in. Every cell phone and laptop has one of these.

Inverters

An inverter is a power supply that converts DC into AC. A common unit converts 12 volts from a car battery into standard 120 volt 60 Hz AC. With an inverter you can operate common household lights and appliances from a battery.

A special variation of an inverter is the uninterruptible power supply (UPS). These are widely used with computers and networks to provide power even during loss of AC input from the power line. A battery operates an inverter that is used to power the computers and other equipment. If the AC power fails, the equipment continues to operate. The battery is continuously charged by a built-in battery charger that operates from the AC power line.

Solar Power System

Any home or business solar power system is similar to a large UPS (see Figure 4.23). It uses the solar panels to charge a bank of batteries that store the energy for use later when the sun goes down. The inverters convert the DC into standard 120/240 volt AC to power the house. In some cases, unused power is sent back to the power grid.

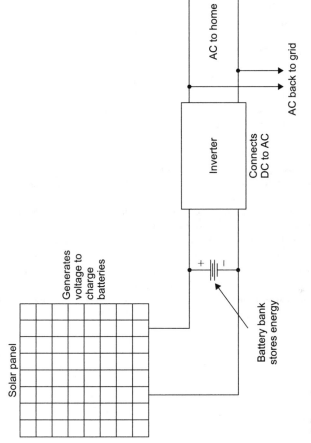

FIGURE 4.23   Common DC-to-AC solar power system.

Project 4.1

Build a Kit

If you are really serious about electronics, then one of the first things you should do is build a kit. A kit, in case you really do not know, is a complete package of electronic parts to make a specific product. You build it yourself by placing components on a printed circuit board (PCB) and soldering them in place. There is no better way to see how electronic products get put together than to handle the wide variety of parts. It is a great learning experience.

There are two main questions to answer about this project. First, what kind of kit do you want? Second, can you learn to solder? You will have to answer the first question, but as for the second, yes, you can learn to solder. It's not that hard.

One way to get an overview of the kits available is to obtain catalogs from the sources mentioned earlier in this book. You can order a print copy online and see the whole range of kits available. Go to:

All Electronics: www.allelectronics.com

Elenco Electronics: www.elenco.com

Jameco: www.jameco.com

Kelvin: www.kelvin.com

Ramsey: www.ramseyelectronics.com

What I recommend is to get a kit related to one of the topics in this chapter. A power supply is a great choice, and it will give you a variable DC voltage source that you can use in future projects. The same is true for a basic audio amplifier. A solar panel kit is a good one also. Most of the companies have lots of radio- and wireless-related kits, but save these for the later chapters on wireless products.

Don't forget to get a small soldering iron (25 watts or so) with a small tip and some solder. You will also need a few hand tools such as a wire stripper, needle-nose pliers, side cutters, and a small screwdriver, if you do not already own these.

Chapter 5

Electronic Circuits: Digital

Practically Everything Is Digital These Days

In this Chapter:
- Binary number system.
- Logic gates and flip-flops.
- Common combinational and sequential circuits.
- Semiconductor memory.
- Programmable logic devices.
- Analog-to-digital and digital-to-analog conversion.

INTRODUCTION

Recall that there are two basic types of electronic signals—analog and digital. A digital signal is one that varies in discrete steps. Unlike an analog signal, which varies continuously, a digital signal has two levels or states. The signal switches or changes abruptly from one state to the other.

Figure 5.1 shows a DC digital signal that switches between two known levels such as zero volts or close to it (<0.1 volt) or 0 V and +3.3 V. The positive voltage can be anything between about 1 volt and 12 volts with 3.3 and 5 being the most common.

Digital signals with two discrete levels are also referred to as *binary signals*. *Binary* means two—two states or two discrete levels of voltage.

Humans use the decimal number system that represents quantities with digits 0 through 9. However, digital equipment and computers do not. Internally, digital equipment processes binary data.

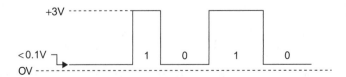

FIGURE 5.1   Binary signal that represents 0 and 1 as voltage levels.

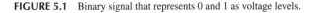

doi: 10.1016/B978-1-85617-700-9.00005-9

BINARY NUMBERS

The binary number system is a set of rules and procedures for representing and processing numerical quantities in digital equipment. Because the base of the binary number system is 2, only two symbols (0 and 1) are used to represent any quantity. The symbols 0 and 1 are called *binary digits* or *bits*. For example, the 6-bit number 101101 represents the decimal number 45. Your understanding of how digital circuits, microprocessors, and related equipment process data is tied directly to an understanding of the binary number system.

The reason for using binary numbers in digital equipment is the ease with which they can be implemented. The electronic components and circuits used to represent and process binary data must be capable of assuming two discrete states to represent 0 and 1. Examples of two-state components are switches and transistors. When a switch is closed or on, it can represent a binary 1. When the switch is open or off, it can represent a binary 0. A conducting transistor may represent a 1, whereas a cut-off transistor may represent a 0. The representation may also be voltage levels. For example, a binary 1 may be represented by $+3.3\,V$ and a binary 0 by $0\,V$ as previously shown in Figure 5.1.

Binary-to-Decimal Conversions

The binary system is similar to the decimal system in that the position of a digit in a number determines its weight. Recall that in the decimal system the weights are powers of 10. The right-most digit is units or 1's, and then 10's, 100's, 1000's, and so on, moving from right to left from one digit to the next. The position weights of a binary number are powers of 2:

$2^0 = 1$
$2^1 = 2$
$2^2 = 4$
$2^3 = 8$
$2^4 = 16$
$2^5 = 32$
$2^6 = 64$
$2^7 = 128$
$2^8 = 256$

The position weights of an 8-bit binary number follow:

Bit position	2^7	2^6	2^5	2^4	2^3	2^2	2^1	2^0
Weight	128 (MSB)	64	32	16	8	4	2	1 (LSB)

The most significant digit or bit (MSB) and the least significant bit (LSB) are identified.

Now let's evaluate the decimal quantity associated with a given binary number, 101101. You simply multiply each bit by its position weight and add the values to get the decimal equivalent, 45.

Position weight	32	16	8	4	2	1
Binary number	1	0	1	1	0	1

$$1 \times 1 = 1$$
$$0 \times 2 = 0$$
$$1 \times 4 = 4$$
$$1 \times 8 = 8$$
$$0 \times 16 = 0$$
$$1 \times 32 = 32$$

Decimal equivalent = 45

You can see that positions with a 0 bit have no effect on the value. Therefore, they can be ignored. To quickly determine the decimal equivalent of a binary number, simply sum the weights of the positions containing a 1 bit. For example, in the number 11101, the weights of those positions with a 1 bit from right to left are $1 + 4 + 8 + 16 = 29$.

Using Hardware to Represent Binary Quantities

Switches are widely used to enter binary data into computers and digital equipment. If the switch is set to the up position, a binary 1 is represented. If the switch is down, a binary 0 is represented. The switch contacts can be open or closed depending on the circuit in which the switch is connected. Figure 5.2 shows a group of slide switches set to represent a binary number, 11000101 or 197.

Indicator lights such as light-emitting diodes (LEDs) are sometimes used to read out binary data in digital equipment. An "on" light is a binary 1 and an "off" light is a binary 0. See Figure 5.3. The decimal value being represented by the display is 178.

Decimal-to-Binary Conversions

You may find it necessary to convert decimal numbers into their binary equivalents. This can be done by dividing the decimal number by 2, dividing the quotient by 2, dividing that quotient by 2, and so on, recording the remainders until the quotient is zero. The remainders form the binary equivalent.

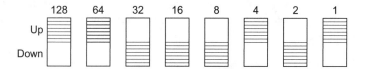

FIGURE 5.2   Off/on switches are used to enter binary data into digital circuits.

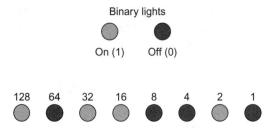

FIGURE 5.3   How lights or LEDs are used to represent binary data.

In the following example, we convert the decimal number 57 to its binary equivalent.

Quotient	Remainder
$57 \div 2 = 28$	1 (LSB)
$28 \div 2 = 14$	0
$14 \div 2 = 7$	0
$7 \div 2 = 3$	1
$3 \div 2 = 1$	1
$1 \div 2 = 0$	1 (MSB)

You can always check your work by reconverting the binary number back to decimal using the procedure described earlier. Also, keep in mind that some scientific calculators can do binary-to-decimal and decimal-to-binary conversions.

Another name for a binary number is *binary word*. The term *word* is more general. It refers to a fixed group of bits that can mean numbers, letters, or special characters and codes. We say that digital equipment processes *binary data words*.

All digital circuits and microcomputers work with a fixed-length binary word. A common binary word length in microcomputers and other digital equipment is 8 bits. All data storage, processing, manipulation, and transmission are carried out in 8-bit groups. Word lengths of 4, 8, 12, 16, 32, and 64 bits are common in digital equipment.

An 8-bit word is usually called a *byte*. In data communication applications, an 8-bit word is sometimes called an *octet*. A 4-bit word is sometimes referred to as a *nibble*.

Maximum Decimal Value for N Bits

The number of bits in a binary word determines the maximum decimal value that can be represented by that word. This maximum value is determined with the formula:

$$M = 2^N - 1$$

where M is the maximum decimal value and N is the number of bits in the word.

For example, what is the largest decimal number that can be represented by 4 bits?

$$M = 2^N - 1 = 16 - 1 = 15$$

With 4 bits, the maximum possible number is binary 1111, or decimal 15.

The maximum decimal number that can be represented with 1 byte is 255 or 11111111. An 8-bit word greatly restricts the range of numbers that can be accommodated. But this is usually overcome by using larger words.

There is one important point to know before you leave this subject. The formula $M = 2^N - 1$ determines the maximum decimal quantity (M) that can be represented with a binary word of N bits. This value is 1 less than the maximum number of values that can be represented. The maximum number of values that can be represented (Q) is determined with the formula:

$$Q = 2^N$$

With 8 bits, the maximum number of values is 256 or 0 through 255.

Table 5.1 gives the number of bits in a binary number and the maximum number of states that can be represented.

BCD and ASCII

The binary numbers we have been discussing are usually referred to as pure binary codes. But there are other types of binary codes. For example, the binary-coded decimal (BCD) system is popular. BCD is a hybrid code between the binary and decimal systems. It was developed in an attempt to simplify the conversion processes between the two systems and to improve human–machine communication.

TABLE 5.1 Number of Bits in Binary Number and Maximum Number of States

Number of Bits	Maximum States
8	256
12	4096 (4 K)
16	65,536 (64 K)
20	1,048,576 (1 M)
24	16,777,216 (16 M)
32	4,294,967,296 (4 G)

Note: K = kilobits = 1024; M = megabits = 1,048,576; G = gigabits = 1,0737,41,824.

To represent a decimal number in the BCD system, each digit is replaced by its 4-bit binary equivalent. Thus, the number 729 in BCD is:

7 2 9

0111 0010 1001

The BCD code is given in Table 5.2.

It is important to note that the 4-bit binary numbers 1010 through 1111 that represent the decimal values 10 through 15 are invalid in BCD.

To convert a BCD number to decimal, simply substitute the decimal equivalent of each 4-bit group. The BCD number 1001 0100 0110 in decimal is 946.

A special form of BCD code is used in computers and data communications systems. It is a 7- or 8-bit code that is used to represent not only numbers but also letters (both uppercase and lowercase), special symbols, and control functions. This code is called the American Standard Code for Information Interchange, or ASCII (pronounced "ask-key").

The following are examples of ASCII:

Number, letter, or symbol	ASCII
F	01000110
8	00111000
J	01101010
+	00101011

TABLE 5.2 Binary-Coded Decimal

Decimal Value	BCD Code
0	0000
1	0001
2	0010
3	0011
4	0100
5	0101
6	0110
7	0111
8	1000
9	1001

The complete ASCII is given in Table 5.3. It provides the decimal and hexadecimal (hex) values of the codes. The code contains both uppercase and

TABLE 5.3 The ASCII Character Code

Char	Dec	Oct	Hex	Char	Dec	Oct	Hex	Char	Dec	Oct	Hex	Char	Dec	Oct	Hex
(nul)	0	0000	0x00	(sp)	32	0040	0x20	@	64	0100	0x40	`	96	0140	0x60
(soh)	1	0001	0x01	!	33	0041	0x21	A	65	0101	0x41	a	97	0141	0x61
(stx)	2	0002	0x02	"	34	0042	0x22	B	66	0102	0x42	b	98	0142	0x62
(etx)	3	0003	0x03	#	35	0043	0x23	C	67	0103	0x43	c	99	0143	0x63
(eot)	4	0004	0x04	$	36	0044	0x24	D	68	0104	0x44	d	100	0144	0x64
(enq)	5	0005	0x05	%	37	0045	0x25	E	69	0105	0x45	e	101	0145	0x65
(ack)	6	0006	0x06	&	38	0046	0x26	F	70	0106	0x46	f	102	0146	0x66
(bel)	7	0007	0x07	'	39	0047	0x27	G	71	0107	0x47	g	103	0147	0x67
(bs)	8	0010	0x08	(	40	0050	0x28	H	72	0110	0x48	h	104	0150	0x68
(ht)	9	0011	0x09	)	41	0051	0x29	I	73	0111	0x49	i	105	0151	0x69
(nl)	10	0012	0x0a	*	42	0052	0x2a	J	74	0112	0x4a	j	106	0152	0x6a
(vt)	11	0013	0x0b	+	43	0053	0x2b	K	75	0113	0x4b	k	107	0153	0x6b
(np)	12	0014	0x0c	,	44	0054	0x2c	L	76	0114	0x4c	l	108	0154	0x6c
(cr)	13	0015	0x0d	-	45	0055	0x2d	M	77	0115	0x4d	m	109	0155	0x6d
(so)	14	0016	0x0e	.	46	0056	0x2e	N	78	0116	0x4e	n	110	0156	0x6e
(si)	15	0017	0x0f	/	47	0057	0x2f	O	79	0117	0x4f	o	111	0157	0x6f
(dle)	16	0020	0x10	0	48	0060	0x30	P	80	0120	0x50	p	112	0160	0x70
(dc1)	17	0021	0x11	1	49	0061	0x31	Q	81	0121	0x51	q	113	0161	0x71
(dc2)	18	0022	0x12	2	50	0062	0x32	R	82	0122	0x52	r	114	0162	0x72
(dc3)	19	0023	0x13	3	51	0063	0x33	S	83	0123	0x53	s	115	0163	0x73

(Continued)

TABLE 5.3 (Continued)

Char	Dec	Oct	Hex	Char	Dec	Oct	Hex	Char	Dec	Oct	Hex	Char	Dec	Oct	Hex
(dc4)	20	0024	0x14	4	52	0064	0x34	T	84	0124	0x54	t	116	0164	0x74
(nak)	21	0025	0x15	5	53	0065	0x35	U	85	0125	0x55	u	117	0165	0x75
(syn)	22	0026	0x16	6	54	0066	0x36	V	86	0126	0x56	v	118	0166	0x76
(etb)	23	0027	0x17	7	55	0067	0x37	W	87	0127	0x57	w	119	0167	0x77
(can)	24	0030	0x18	8	56	0070	0x38	X	88	0130	0x58	x	120	0170	0x78
(em)	25	0031	0x19	9	57	0071	0x39	Y	89	0131	0x59	y	121	0171	0x79
(sub)	26	0032	0x1a	:	58	0072	0x3a	Z	90	0132	0x5a	z	122	0172	0x7a
(esc)	27	0033	0x1b	;	59	0073	0x3b	[	91	0133	0x5b	{	123	0173	0x7b
(fs)	28	0034	0x1c	<	60	0074	0x3c	\	92	0134	0x5c	|	124	0174	0x7c
(gs)	29	0035	0x1d	=	61	0075	0x3d	]	93	0135	0x5d	}	125	0175	0x7d
(rs)	30	0036	0x1e	>	62	0076	0x3e	^	94	0136	0x5e	~	126	0176	0x7e
(us)	31	0037	0x1f	?	63	0077	0x3f	_	95	0137	0x5f	(del)	127	0177	0x7f

Char = ASCII character
Oct = Octal (3-bit groups)
Hex = Hexadecimal (4-bit groups)
Dec = Decimal
Characters in parentheses are control codes for specific operations.
X = "Don't care" character
Example: ASCII capital letter A
  Decimal 65
  Binary 00000000001000001
  Octal 0/000/000/001/000/001
          0 0 1 0 1
  Hex 0000/0000/0100/0001
          0 x 4 1

lowercase letters, the digits 0 through 9, punctuation marks, common mathematical symbols, and many special two- and three-letter codes. The special codes are known as control codes. They are used to indicate the start and end of a string of characters or to initiate special operations. Some examples follow:

STX—Start of text sequence
EOT—End of transmission of a string of characters
BEL—Ring a bell
DEL—Delete
LF—Line feed
CR—Carriage return

These codes not only represent alphanumeric data but can also make things happen in a printer or computer.

The main use of ASCII is in data communication. Computers use ASCII to "talk" to their peripheral units, such as printers, or to one another. A common example is a personal computer that receives data via a modem from the Internet. Another example is the data sent from a computer to a laser printer. ASCII is the communication code for text.

Hexadecimal Notation

Hexadecimal notation is a way to represent binary values with numbers and letters in an effort to make data easier to remember, and to reduce the errors normally associated with writing or reading long strings of 1's and 0's. The hex code is just an extension of the BCD code, with the decimal values from 11 to 15 represented by the letters A through F. See the following table.

Decimal Value	Hex Code
0	0000
1	0001
2	0010
3	0011
4	0100
5	0101
6	0110
7	0111
8	1000
9	1001
A (10)	1010
B (11)	1011
C (12)	1100
D (13)	1101
E (14)	1110
F (15)	1111

(Continued)

Hexadecimal Notation (Continued)

As an example to convert the binary number 101000110101100 to hex, you divide the word into 4-bit groups starting on the right and then replacing each with the hex codes. In this case, the hex code is A35C.

Parallel and Serial Data

It is frequently necessary for digital devices to communicate. Someone using a computer may have to send data to a printer. Or two people using personal computers may have to communicate over a local area network (LAN). To accomplish either of these objectives, one must transmit binary data from one circuit to another. This is done in one of two ways: parallel transmission or serial transmission.

Parallel Data

When the parallel method is used, all bits of a binary word or number are transmitted simultaneously. Figure 5.4 shows an 8-bit word being transferred from logic circuit 1 to logic circuit 2.

A *bus* is a set of parallel data lines over which digital data is transferred. A data bus can have any number of data lines as required by the application. However, the main characteristic of a bus is parallel data transfer. In physical form, a bus is usually implemented on a connector, with lines of copper on a printed circuit board or with a ribbon cable with one wire per bit.

Because all bits are transmitted at the same time over a parallel bus, data movement is extremely fast. The main benefit of parallel data transmission is its high speed. With high-speed logic circuits, a binary word can be transferred from one point to another in as little as several nanoseconds. Parallel data transmission over buses in computers takes place at rates up to several hundred megahertz. It is not uncommon to see a rate of 100 million bits per second (Mbps)

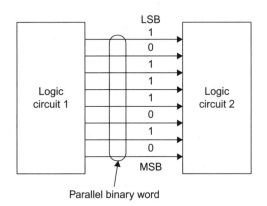

FIGURE 5.4   Parallel binary data transfer over bus.

on a personal computer data bus. On long buses, data rates are slower because the stray inductance and capacitance of the cable typically distorts the data and slows the transmission speed.

Although parallel transmission is fast, it is expensive, because there must be circuitry for each bit on both the sending and receiving ends. There must be one wire per bit in a cable plus a separate ground wire. This increases the complexity and thereby the cost of the circuitry.

Serial Data

The other method of moving binary data from one point to another is serial transmission. In the serial method, the bits of a word are transmitted one at a time, or sequentially. Figure 5.5 shows the waveform of an 8-bit serial binary word. This is the voltage that you would monitor at the single output of the circuit generating the word. It is the waveform that an oscilloscope would display. Each bit occurs for a fixed time interval such as 1.5 milliseconds. Thus, it takes $1.5 \times 8 = 12$ milliseconds (ms) to transmit an 8-bit word.

The speed of data transmission is measured in bits per second (bps). That figure can be calculated from the expression:

$$bps = \frac{1}{t}$$

Here, t is the time for 1 bit. In Figure 5.5, 1 bit takes $1.5\,\mu s$. The speed of the data, then, is

$$bps = 1 \div (1.5 \times 10^{-6}) = 666,666.7\,bps \text{ or } 666.7\,kbps$$

If you know the speed in bps, you can calculate the bit time (t) by rearranging the formula:

$$T = \frac{1}{bps}$$

Assume a data rate of 14,400 bps. The bit time is $69.4\,\mu s$.

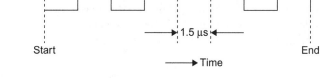

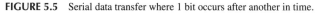

FIGURE 5.5   Serial data transfer where 1 bit occurs after another in time.

As seen in Figure 5.5, the MSB is transmitted first. Depending on the system used, however, the LSB can be sent first. In any case, the speed of transmission depends on the number of bits in the word and the duration of each bit. It is this longer transmission time that is the primary disadvantage of the serial method. However, although serial transmission is slower than parallel transmission, it is perfectly suitable for many applications. Common data rates are in excess of millions of bits per second. Data rates of more than 10 Gbps are common.

The main benefits of the serial method are its simplicity and low cost. Only a single line rather than multiple lines is needed to interconnect the equipment. In addition, only one set of sending and receiving circuits is needed. All data transmissions by copper cable or fiber optical cable exceeding about 100 Mbps are made by the serial method. All wireless data transmissions are serial.

DIGITAL LOGIC ELEMENTS

Digital logic elements are the basic circuits that are used to process the binary data. The logic element has one or more binary data inputs to be processed. The logic element processes or manipulates the binary input signals in a fixed way and generates an appropriate output signal. The output is a function of the binary states of the inputs and the unique processing capability of the logic element. The logic element "looks" at the binary input signals, and then makes a decision and generates an appropriate output. There are five elementary logic circuits: inverter or NOT, AND, OR, NAND, and NOR. These functions are performed by logic elements known as inverters and gates.

You need not be concerned with the inner workings of these logic elements. Instead, your primary concern is with the basic logic functions. Needless to say, the gates and inverters are inside every microcomputer and digital integrated circuit.

Inverter

The inverter has a single input and a single output. The logic function performed by an inverter is *inversion*. The output of the inverter is simply the inverse, or opposite, of the input. Because binary signals can assume only one of two different states, either 0 or 1, the inverter generates a 0 when the input is a 1, and a 1 when the input is a 0.

Two common symbols used to represent an inverter are shown in Figure 5.6. The input and output are given names, which are usually letters of the alphabet.

There are a variety of ways to express the operation of various logic functions. One way is to use Boolean algebraic expressions. A Boolean expression is a simple mathematical or algebraic formula that expresses the output in terms of the input. The Boolean expression for the inverter is:

$$B = A^*$$

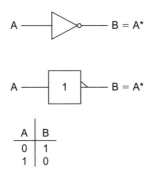

FIGURE 5.6   Logic symbols for inverter and truth table.

This equation is read, "B equals not A." The asterisk means NOT or inversion. What this expression is telling you is that if the input is A, the output, B, is not A.

Another method of expressing the function of a logic element is a truth table. This is simply a table listing all possible combinations of the inputs and outputs of a logic element. Figure 5.6 also shows the truth table for an inverter. There are only two possible inputs, 0 and 1, and the corresponding outputs. A truth table completely defines the operation of a logic element.

AND Gate

A gate is a circuit that has two or more inputs and a single output. The gate generates a binary output that is a function of the states of the inputs and the nature of the gate itself.

The type of gate determines how the binary inputs are processed. The two basic logic gates are the AND gate and the OR gate. An AND gate is a logic circuit that generates a binary 1 output if all of its inputs are binary 1's. Otherwise, the circuit generates a binary 0. All of the inputs have to be binary 1 for the AND gated to produce a binary 1 output. If any one of the inputs is a binary 0, the output is binary 0.

Figure 5.7 shows the logic symbols used to represent an AND gate. Letters of the alphabet are used to represent inputs and outputs. There may be more than two inputs, but there is only one output.

As with the inverter, the output of the AND gate can be expressed in terms of the inputs with a Boolean expression. The Boolean expression for an AND gate is:

$$C = AB$$

This equation is read, "C equals A and B." Parentheses also can be used to separate the inputs; for example, $C = (A)(B)$.

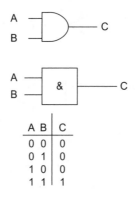

A B	C
0 0	0
0 1	0
1 0	0
1 1	1

FIGURE 5.7   Logic symbols for AND gate and truth table.

The truth table for a simple two-input AND gate is also shown in Figure 5.7. Note that the output C is a binary 1 when both inputs are binary 1's. At all other times, the output is binary 0. With two inputs, there are a total of four different input combinations ($2^2 = 4$). With three inputs, there are $2^3 = 8$ possible input combinations for an AND gate. These inputs are the binary numbers 000 through 111.

One of the most common applications of an AND gate in digital circuits is gating. Gating simply refers to the use of one binary signal to control another. A two-input AND gate is most often used as a control gate. One input signal is the control that either keeps the other input signal from passing through to the output or allows it to pass.

Figure 5.8 shows the inputs and output of a typical control gate. Note that multiletter names called *mnemonics* are used instead of single letters to designate the input and output signals. The ENB or enable input represents the control signal. As long as the ENB input is a binary 0, the output is a binary 0. The gate is said to be inhibited because nothing more than a binary 0 output occurs.

When the ENB control input is a binary 1, the gate is enabled. At this time, the main input signal, CLK for clock, is allowed to pass through to the output BST. CLK is a periodic clock signal that switches repetitively between the binary 0 and the binary 1 levels at a fixed frequency. The output of the AND gate follows the CLK input as long as the ENB control input is a binary 1.

OR Gate

Another commonly used logic gate is the OR circuit. Like the AND gate, the OR gate has two or more inputs and a single output. The OR gate generates a binary 1 if any one or more of its inputs are binary 1. The only time the output of an OR gate is binary 0 is when all its inputs are binary 0.

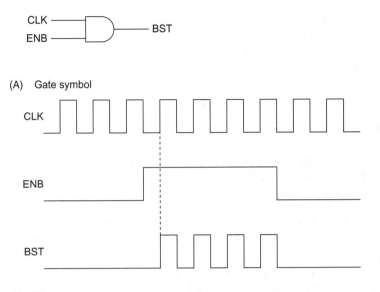

(A)   Gate symbol

(B)   Waveforms

FIGURE 5.8   How AND gate is used to turn signal off or on. **(A)** The AND gate showing the input and output names. **(B)** The waveforms of the inputs and output.

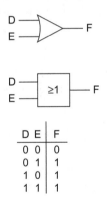

D	E	F
0	0	0
0	1	1
1	0	1
1	1	1

FIGURE 5.9   Logic symbols for OR gate and truth table.

The symbols used to represent an OR gate are shown in Figure 5.9. Letters of the alphabet are used to designate the inputs and outputs. Using these designations, the Boolean output expression for the OR gate is:

$$D = E + F$$

This equation is read, "D equals E or F." In this Boolean expression, the plus sign designates the OR function. In a Boolean logic expression, the plus sign does not mean addition. In Figure 5.9, the mathematical symbol ≥ (greater than or equal to) is used to designate the OR logic function. The truth table for an OR gate further defines its operation. The output D is a binary 1 when input E or input F or both are a binary 1. Like an AND gate, an OR gate can have any number of inputs.

NAND and NOR Gates

The AND and OR logical functions are basic to all digital systems. Although these functions are implemented with AND and OR gates, a variation of these gates is even more widely used. These are NAND and NOR gates, which are a combination of either an AND gate or an OR gate and an inverter. These gates are more flexible in their application when they are used together.

A NAND gate can be made with an AND gate followed by an inverter. Figure 5.10 shows this circuit configuration. Note the two special symbols in Figure 5.10 used to represent the logical NAND function. The truth table is also given.

A NAND gate can perform all of the gating and detection functions mentioned earlier for an AND gate. The only difference is that the output is inverted.

The other widely used digital logic element is a NOR gate. A NOR gate is a combination of an OR gate followed by an inverter (see Figure 5.11). The symbols used to represent the NOR logic function are also shown. The output of a NOR gate is simply the complement of the output of an OR gate, as shown by the truth table. An XOR actually is a 1-bit binary adder.

Exclusive OR Gate

A variation of the OR gate is the exclusive OR, usually designated XOR. Its symbols are given in Figure 5.12. The truth table is also shown. Note that the

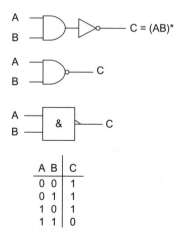

A B	C
0 0	1
0 1	1
1 0	1
1 1	0

FIGURE 5.10 Logic symbols for NAND gate and truth table.

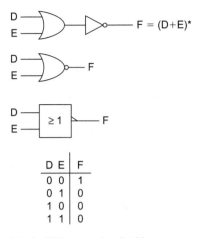

FIGURE 5.11   Logic symbols for NOR gate and truth table.

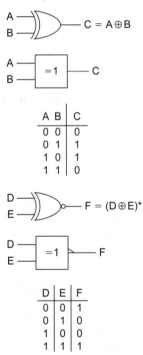

FIGURE 5.12   Logic symbols for XOR and XNOR gates and truth tables.

circuit only produces a 1 output if the two inputs are opposite. The output is 0 if the two inputs are the same.

An inverted XOR or XNOR has the symbols and truth table shown in Figure 5.12. It is useful as a 1-bit comparator since the output is 1 if both inputs are the same.

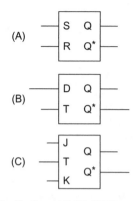

FIGURE 5.13   Logic symbols for flip-flops. **(A)** RS. **(B)** D type. **(C)** JK.

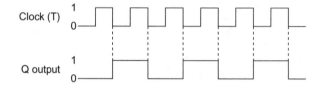

FIGURE 5.14   Toggling T input on JK flip-flop causes it to change states, producing frequency division by 2.

Flip-Flops

A flip-flop (FF) is another basic building block of electronics. It is essentially a circuit that stores 1 bit of data. The circuit has two states—set and reset. If the FF is storing a binary 1, it is set. If it is storing a binary 0, it is reset. Figure 5.13A shows the basic block diagram of a reset-set (R-S) FF. The inputs are set (S) and reset (R). The two outputs Q and Q* are complementary. You apply an input to the appropriate input to set or reset the FF.

A D-type flip-flop is shown in Figure 5.13B. It has a D or data input and a clock (T) input. When the clock occurs, the FF is either set or reset depending on the value of the D input. The D FF is used to store 1 bit of data.

Another type of FF is the JK, shown in Figure 5.13C. It has J and K inputs that correspond to S and R inputs, respectively. It also has a clock or T input that performs the set or reset operations. To store a binary 1, you make J = 1 and K = 0. When a clock input occurs, the FF is then set or reset.

The T input also causes the state of the FF to toggle or change state when a clock transition occurs (see Figure 5.14). This feature is used to produce frequency division by 2 as shown. The output is a square-wave half the frequency of the T input. Additional JK FFs can be cascaded to produce frequency division by the power of 2 or 2, 4, 8, 16, 32, 64, and so on, as shown in Figure 5.15.

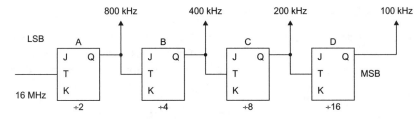

FIGURE 5.15 Cascading JK flip-flops produce higher-frequency division ratios, which is also a binary counter that counts from 0000 to 1111.

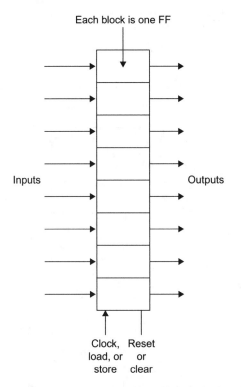

FIGURE 5.16 Storage register from one word consists of FFs. It also has clock/load and clear/reset inputs.

Storage Registers

A storage register is a place to store a binary word or number. It is made up of 1 FF per bit. An 8-bit register stores bytes and has 8 FFs. Figure 5.16 shows a register with eight inputs and eight outputs. Most registers also have a clock input also called LOAD or STORE that determines when the input data is stored. Some also have a CLEAR or RESET input that puts all the FFs in the binary 0 state.

Shift Registers

A shift register (SR) is like a storage register, but the bits stored there may be shifted from one FF to the next as each clock pulse occurs. Figure 5.17A shows how serial binary data is shifted into the SR for storage. As new data

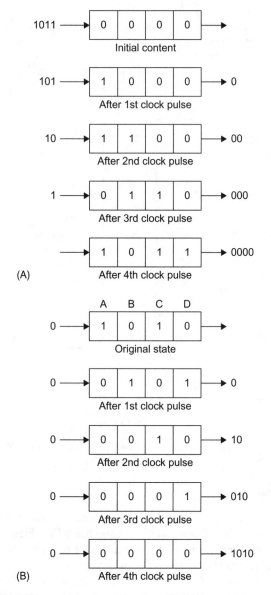

FIGURE 5.17 (A) Shifting serial data into shift register. (B) Shifting serial data out of shift register.

is shifted in, the existing data will be shifted out and sent to another circuit. Figure 5.17B shows data being shifted out as binary 0's are shifted in.

Shift registers are often used for parallel-to-serial and serial-to-parallel conversions. A serial-to-parallel conversion is shown in Figure 5.18A. The SR initially contains all binary 0's. Then serial data is shifted in and the output is taken in parallel from the FF outputs. For parallel-to-serial conversion, a binary word is initially stored in the register. Then the word is shifted out as the clock pulses occur (see Figure 5.18B).

Counters

As its name implies, a counter is a circuit that counts input pulses. Counters are made of multiple cascaded FFs, and they are sometimes accompanied by logic gates. As the input pulses occur, the binary number stored in the counter FFs tells how many pulses have occurred. The frequency divider shown in Figure 5.15 is also a 4-bit counter. It can count from 0 (0000) to 15 (1111).

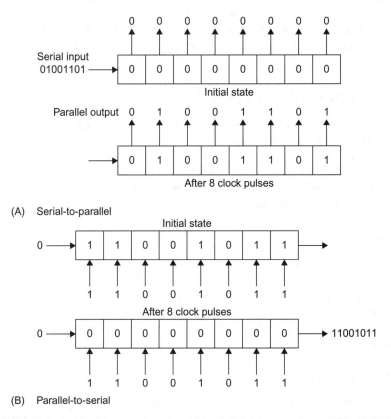

(A)   Serial-to-parallel

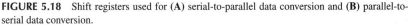

(B)   Parallel-to-serial

FIGURE 5.18   Shift registers used for (A) serial-to-parallel data conversion and (B) parallel-to-serial data conversion.

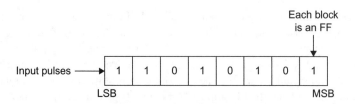

FIGURE 5.19   Counter storing binary value equal to number of input pulses that occurred.

If all FFs are initially reset to 0, then 10 pulses occur, and the FFs from A to D would be storing 0101 from left to right. The right-most FF (D) is the MSB, so you read the counter as 1010 or the binary equivalent of decimal 10.

Binary counters can store numbers up to $2^N - 1$, where N is the number of FFs. For example, an 8-bit counter would count from 00000000 to 11111111 or $2^8 - 1 = 256 - 1 = 255$.

Figure 5.19 shows a counter where 178 pulses have occurred, so the counter stores the number 10101011 that is the binary equivalent of 179.

Another popular form of counter is the BCD version that counts from 0000 to 1001 or 0 through 9. BCD counters are used to keep track of decimal quantities in binary form. BCD counters can serve as frequency dividers as well. Each divides by 10.

COMMON LOGIC CIRCUITS

There are several types of logic circuits that occur again and again in digital equipment. The counters and registers just described are good examples. A few others that you may encounter are described in this section. Remember that these circuits can be made from individual gates and FFs in IC form. You can also buy them in IC form already wired for their specific function. But mostly these circuits appear as a collection of circuits inside some larger IC.

Multiplexer

A multiplexer or mux is a circuit with two or more inputs and one output (see Figure 5.20). Here there are four inputs and one output. Two other inputs are used to select one of the inputs to appear as the output. The four selected inputs are 00, 01, 10, and 11. In the figure, note which input is routed to the output with each input select code. Multiplexers can have as many inputs as needed.

Demultiplexer

A demultiplexer or demux is the opposite of a mux. It has one input and multiple outputs. Like the mux, select inputs specify which of the output lines the input is connected to.

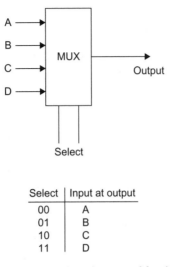

Select	Input at output
00	A
01	B
10	C
11	D

FIGURE 5.20   Digital multiplexer (mux) that selects one of four inputs to appear at the single output.

Decoder

A decoder is a circuit that can detect a specific binary code. In its simplest form, it is just an AND gate, as shown in Figure 5.21A. If the input 0110 appears at the inputs, the AND gate will have all its inputs at binary 1, so the output will be binary 1 indicating that it is detecting the desired input. For any other 4-bit input, the output will be binary 0.

Decoders can also be made to detect all states of a given binary input. For example, a 4-bit decoder like that in Figure 5.21B looks at the 4-bit inputs and decodes all states from 0000 to 1111. If the input is 1000, the 8 output is binary 1 and all other outputs are 0.

Comparators

A comparator takes two binary words as inputs and compares them. If the two inputs are equal, a binary 1 output occurs. Otherwise, if the two inputs are different, the output is 0. Some comparators have outputs that can tell if one input is larger than or smaller than the other.

Arithmetic Circuits

Digital logic does a good job of basic arithmetic functions like add and subtract. The basic computing element is called an arithmetic logic unit (ALU). It does binary addition and subtraction. It can also perform basic logic operations such as AND, OR, or XOR on two binary input words. Circuits for multiplication and division are more complex, but are common in digital computers.

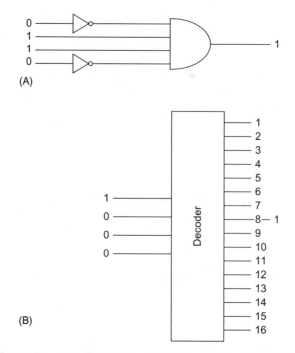

FIGURE 5.21 Decoders. **(A)** 1-AND gate decoder that identifies input 0110. **(B)** 4-to-16 decoder that identifies 1 of 16 different 4-bit input values.

DIGITAL MEMORIES

Digital memories are semiconductor circuits that store binary data. You have probably heard of the terms RAM and ROM as they apply to personal computers. These are the two main types of memory, random access memory (RAM), and read-only memory (ROM).

RAM

The random access memory is one in which you can access hundreds, thousands, or millions of binary storage locations. You can store one binary number or word in each location or read the data out of the location. Each location is referred to by a unique binary number called the address. You send the memory an address and activate the selected location. Then you can read data from the location or write data into that location. RAMs therefore are what we call read-write memories.

Figure 5.22 shows the concept of a RAM. It has eight locations for 4-bit words. To address eight locations you need a 3-bit address ($2^3 = 8$). When you apply the address to the address decoder, the decoder activates the desired location. In the figure you see that the binary number 1001 stored in location 011 is being accessed and read out.

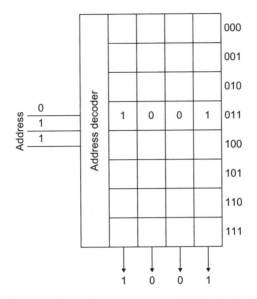

FIGURE 5.22 Basic concept for random access memory (RAM). Address selects one of eight storage locations. Then data may be read from that location or data stored in it.

Most memories are large today. We talk about megabytes (MB) or millions of bytes of data or gigabytes (GB) or billions of bytes of data in a modern PC. Embedded controllers will have fewer locations, usually in the thousands of bytes (KB) or perhaps more in the larger systems. Storage word sizes are usually 8 (bytes), 16, 32, or 64 bits.

There are two ways to make a RAM. The first way is just to build one FF for each bit to be stored. The basic set-reset FF described earlier is the basic storage circuit (see Figure 5.23A). Memories using this type of circuit are called static RAM or SRAM. They are called static because once you store the bit in the FF, it stays there as long as power is applied to the circuit.

The other form of RAM is called *dynamic RAM* or DRAM. The basic storage circuit is a capacitor as shown in Figure 5.23B. If you charge the capacitor, it is storing a binary 1. If the capacitor is discharged, a binary 0 is being stored. A MOSFET is used as a switch to access the capacitor. If the MOSFET is turned on by an input address on the gate, it connects the capacitor to the data line, where it may be read out or where a new bit may be stored. When the MOSFET is turned off, the capacitor is isolated and just stores the bit.

The reason why the term *dynamic* is used for this type of RAM is that the charge on the capacitor will leak off very quickly. The capacitor is very small and millions of them are made on a tiny chip of silicon. For this reason, the capacitor has to be recharged every so often so that the data is retained. This is called a refresh operation and it usually occurs every 2 milliseconds or so.

Comparing SRAM and DRAM, you should know that SRAM is the faster of the two in that you can store data or read data out faster than you can with a

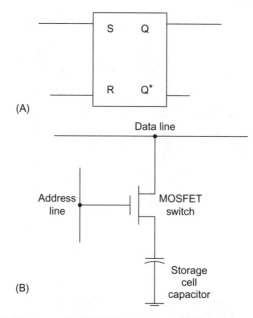

FIGURE 5.23   RAM storage cells. **(A)** RS FF for SRAM. **(B)** Capacitor cell for DRAM.

DRAM. The downside is that the SRAM is more complex and takes up more space on the chip, and thus costs more. The DRAM is slower but not much. Its advantage is that you can put more storage cells on a chip, making them cheaper. The need to refresh is a downside but easy to live with, as all the refresh circuitry is on the chip and the process is transparent to the user.

Just keep in mind a key fact. RAM is fast and reliable, but it loses all data if the DC power to the memory circuit is removed. All data is erased. RAMs are therefore said to be volatile.

ROM

Read-only memories are nonvolatile memories. Once you store data in them, they retain the data when power is removed. In some ROMs, you can change the data stored there. Incidentally, all ROMs are random access in that you can go to any desired location to read or maybe write data.

There are several categories of ROM. The older types are called masked ROM, where the mask refers to the actual wiring of the ROM to permanently fix the data in the circuitry when it is made. The storage cells are essentially a MOSFET matrix where each MOSFET is permanently turned on or off to store 1 or 0. No changes are possible. Later a programmable ROM or PROM was developed. It used a special MOSFET that could be programmed on or off. You could then erase the programming by shining ultraviolet (UV) light

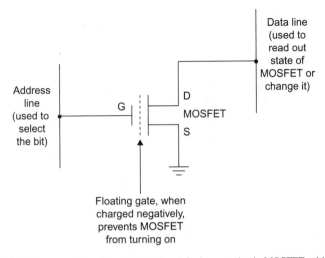

FIGURE 5.24 Storage cell used in EEPROM and flash memories is MOSFET with a special floating gate that keeps transistor off or on.

on the chip for a short time. Then the PROM could be reprogrammed. These are no longer used.

Most ROMs today are what we call electrically erasable programmable ROMs or EEPROMs or E^{2}PROMs. The storage element is a MOSFET that you turn off or on with a programmable voltage (see Figure 5.24). The MOSFET retains that state until it is electrically changed.

The main problem with EEPROMs is that they are slow to erase and reprogram. But today we have a special version of EEPROMs called flash memories. These are also EEPROMs, but they actually are almost as fast as RAMs for reading and writing data. Yet they are nonvolatile.

You will hear of two types of flash memory—NOR and NAND. NOR flash is used to store programs for microcontrollers. It can only be erased on large block sizes and not individual cells. NAND flash is faster and you can erase individual memory locations. The NAND flash is gradually replacing the NOR flash in many applications.

Flash memory has become so popular it is used in almost every electronic product today. MP3 players and iPods store music in flash memory. Cell phones store phone numbers, digital cameras store photos, and in embedded controllers, the flash usually stores the program that tells the processor what to do. And, let's not forget the flash drive or "thumb" drive as it is called. These devices with a USB connector can store many gigabytes of data and are very inexpensive. Flash memory is also used to make solid-state drives (SSDs) that are the equivalent of the hard disk drives used in PCs and laptops. They are still expensive, but as prices decline, you can look for more SSDs to replace conventional mechanical/magnetic hard drives.

PROGRAMMABLE LOGIC DEVICES

There are two ways that digital devices are built today. The most common is to use an embedded controller or microcomputer. This is a single-chip computer that you can program to do whatever digital function you want. Most electronic products have at least one embedded controller at the heart of their design.

In some cases, the embedded controller is not fast enough to do the job. One way to make the circuitry is to use individual logic circuits such as gates, flip-flops (FFs), counters, and so on. That is the way it used to be done. Now, if you need fast custom digital logic, you build the circuit with a programmable logic device (PLD). A PLD is an IC that contains a mix of gates, FFs, and other circuits that can be programmed to implement any digital function. The programming is usually done on a computer and a file is created. The file is then downloaded to the PLD, where it is stored in a RAM or ROM. The file in the memory determines how the unique circuit is formed.

There are three basic types of PLDs: simple PLDs, complex PLDs, and field programmable gate arrays (FPGAs). The SPLDs are small devices for simpler digital circuits. The most common types are called programmable array logic (PAL), programmable logic arrays (PLAs), and generic array logic (GAL). PROMs are also used as SPLDs. For example, the address input to a ROM is some binary number or a collection of binary inputs. These form an address to locate a specific binary word stored in that location. When that particular selection of input bits occurs, it forms the address that identifies the memory location where the desired set of output responses is stored. This form of SPLD is also called a look-up table or LUT.

The other SPLDs are essentially a set of AND and OR gates that can be programmed. They all take the form shown in Figure 5.25. You can select which inputs go to the AND gates, and/or select which AND outputs go to which OR gate inputs. A PROM has the AND array fixed and you can program the OR gates. A PAL is an SPLD in which you can program the AND inputs but the OR connections are fixed. A PLA is a PLD that lets you program both the AND inputs and the OR connections. In both PLAs and PALs, the programming is usually permanent.

A GAL is a variation of the PAL with its AND inputs programmable. The difference is that the programming can be changed. The program you develop is actually stored in flash memory. The flash memory cells send the programming bits to the various AND and OR gates telling them what logic to implement. You can then erase it and reprogram if you want or need to. GALs also have one or more FFs associated with the logic gates, so you can store bits or actually build counters and registers by programming.

CPLDs are complex PLDs similar to SPLDs, only larger. They contain more gates and flip-flops so that very large logic projects can be built on a single chip. A typical CPLD is really just a large collection of PALs or GALs that can be individually programmed and then connected together with a large

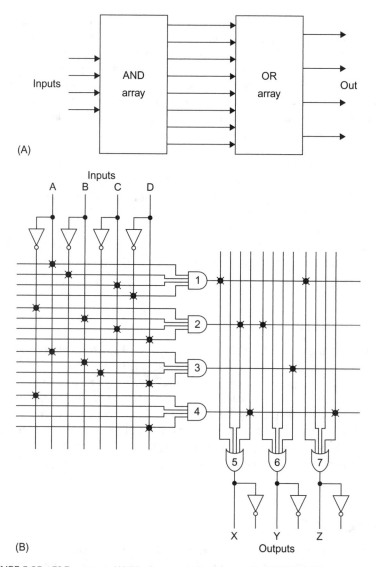

FIGURE 5.25   PLD concept. (**A**) Block concept of programmable AND and OR gate arrays. (**B**) Simple circuit of PLD showing programming by dots on AND and OR gate arrays.

interconnect matrix. Input/output (I/O) circuits also provide inversion if needed or extra power to drive external circuits. Again, flash memory is used to store the programming.

The largest and most flexible PLD is the field programmable logic array or FPGA. It has become the most popular and common way to implement digital equipment other than a microcontroller. FPGAs are large chips that contain thousands of logic circuits. The largest devices have millions of gates and other circuits. Most FPGAs are programmed by storing a bit pattern in the SRAM cells inside. These cells tell the logic circuits what configuration to take and what functions to perform. If you turn the power off, the programming goes away and all you have is a dumb FPGA with no logic interconnections or functions. When you power up an FPGA, the desired program, usually from a flash memory, or a computer loads the RAM cells with the program, thereby enabling the FPGA to perform its desired function.

The logic inside an FPGA is essentially thousands of look-up tables (LUTs) that can be individually programmed. Each LUT has its own program memory block as well as one or more FFs and I/O circuits. The LUTs may also be interconnected to one another by programming. With this arrangement virtually any digital logic operation can be programmed. You can even make your own custom computer with an FPGA, for example.

There are about a half-dozen companies that make PLDs. The two oldest and largest are Xilinx and Altera. Both have extensive lines of CPLDs and FPGAs.

DATA CONVERSION

The term *data conversion* refers to the translation of analog signals into digital form and vice versa. While it used to be an all-analog electronics world, today it is virtually all digital. We still have analog signals such as voice, video, radio waves, and sensor voltages, but mostly we process, store, and transmit data by digital means. For that reason, circuits that perform these conversions are critical to virtually all electronic applications. All cell phone voice is digital, all modern TV is digital, and sensor outputs are captured and processed in digital form. It is difficult to name a product today that does not use some form of data conversion.

Analog-to-Digital Conversion

The process of converting an analog signal into a digital one is called analog-to-digital conversion, and it is performed by an analog-to-digital converter (ADC). The process, also referred to as sampling, is illustrated in Figure 5.26. The ADC looks at the analog input and periodically takes a sample of the voltage at that instant, captures it, and then converts it into a proportional binary number. We say that we are digitizing the signal. The sample points are shown by the dots on the analog curve. The binary value of the sample is shown to the right of the curve. The conversion process actually results in a sequence of binary numbers that represent the analog waveform. These values are usually stored in a RAM or transmitted to other circuits as shown in Figure 5.27. Note the symbol for an ADC.

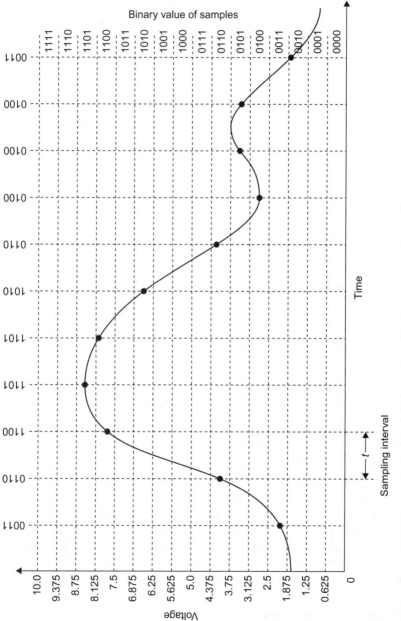

FIGURE 5.26 Analog-to-digital conversion is performed by sampling analog signal at equal intervals and generating proportional binary value.

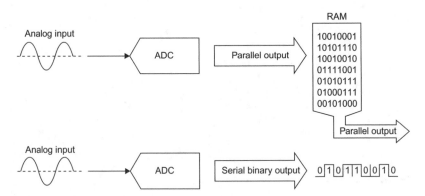

FIGURE 5.27   Symbol for ADC with both parallel and serial output examples.

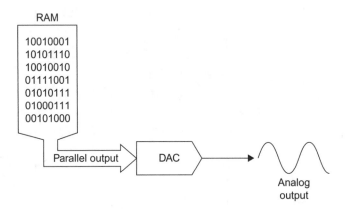

FIGURE 5.28   Converting binary data back into analog signal with DAC.

Digital-to-Analog Conversion

To recover the original signal, we put the data sequence previously captured by the ADC into a digital-to-analog-converter (DAC) (see Figure 5.28). The output is a version of the analog signal. The DAC output is not a perfect reproduction, but just an approximation. This is shown in more detail in Figure 5.29. Each binary input results in a constant voltage output from the DAC during the sample period. The result is a stepped approximation to the original signal. The rate at which the binary data is sent to the DAC must be the same as the sampling interval to recover the original frequency information in the signal.

Resolution and Sampling Interval

The key to good data conversion is to use greater resolution and faster sampling rates. Resolution refers to the number of bits used in the data conversion.

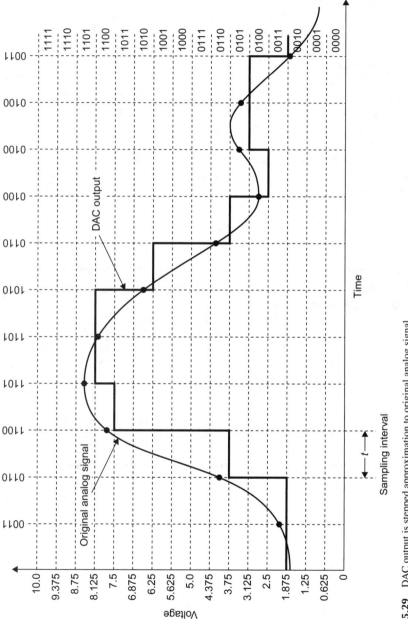

FIGURE 5.29   DAC output is stepped approximation to original analog signal.

In Figure 5.27, only 4 bits are used, so the resolution is poor. The voltage range is only divided into 16 intervals, meaning that amplitude variations at less than 0.625 volt are missed. This problem can be corrected by using more bits. ADCs are available in many bit sizes. The most common are 8, 10, and 12 bits, but 14 and 16 bits are available. Some methods of ADC produce resolutions of 20 to 26 bits. The result is a finer conversion of amplitude detail. As an example, if the 0- to 10-volt range in Figure 5.27 was sampled with a 12-bit ADC, the individual smallest voltage increment that can be detected is $10/2^{12} = 10/4096 = 2.44\,mV$ instead of the 0.625 volt in the figure.

Another critical specification is sampling rate. To retain all the frequency detail in a signal, the sampling rate must be at least twice the highest frequency in the signal. This is called the Nyquist criterion. For example, when digitizing music with a frequency range of 20 Hz to 20 kHz, the sampling rate must be at least double the 20-kHz frequency. In most systems, a rate of 44.1 kHz or 48 kHz is used.

ADCs with sampling rates to several hundred megahertz are commonly available, meaning that even radio signals can be digitized.

Project 5.1
_____

Digging Deeper into Digital

If you want to learn more about digital circuits without investing in one or more new books, you can find some good material online. A few alternatives follow.

Do your own search on digital logic using the terms *tutorial, introduction to, fundamentals,* or *basics.* You will get lots of hits that you can sort through to find what you want.

Go to www.allaboutcircuits.com. Look at their digital section called Volume IV.

Go to openbookproject.net/electricCircuits. Look at the digital chapter.

On both of the websites, there is also a projects or experiments section that will guide you to further hands-on exploration.

Chapter 6

How Microcomputers Work

The Brains of Every Electronic Product Today

In this Chapter:

- Microprocessors, microcomputers, and embedded controllers defined.
- The stored program concept.
- Central processors and how they work.
- Types of microcomputers.
- Modern examples of embedded controllers.
- Digital signal processing.

INTRODUCTION

Whenever you hear or see the word *microcomputer*, you probably think of a personal computer. A personal computer is, of course, one type of microcomputer. But you may not be aware of other forms of microcomputers. For example, do you know that almost every piece of electronic equipment you own or use has a microcomputer in it? This type of microcomputer is called an embedded controller. It is usually a single integrated circuit that performs all of the basic functions of a computer but is dedicated to a specific task. Such micros (*micro* is short for microcomputer) are in TV sets, stereo receivers, microwave ovens, CD players, DVD players, cell phones, blenders, copiers, and most other types of electronic devices. You will find them in your car (several of them), your iPod, your bathroom scale, gasoline pumps, and dozens of other things you use every day. Microcomputers do indeed make your day—you cannot live without them.

CONCEPTS AND DEFINITIONS

A microcomputer is a small digital computer that can take several different forms. It can be a single integrated circuit, or it can be a module made up of several integrated circuits on a printed-circuit board.

doi: 10.1016/B978-1-85617-700-9.00006-0

The two most common types of microcomputers are the embedded micro-controller and the personal computer. One special type of microcomputer is the programmable logic controller (PLC) used for industrial control (Chapter 12).

The larger microcomputers, such as personal computers, are made with a microprocessor. A microprocessor is a large-scale integrated circuit that contains most of the digital logic circuitry usually associated with a digital computer. This logic circuitry is referred to as the central processing unit (CPU). A microprocessor is a single-chip CPU.

Another common micro is called a *core*. A core is a microprocessor or microcomputer integrated with some other circuits. The core is made on the same silicon chip as a cell phone or a security device. Today multicore micros are common. Two or more cores are used to get more processing power.

Embedded microcontrollers are complete micros on a single chip, including the CPU, memory, and input/output circuits. But in all cases, a microcomputer is an assembly of digital logic circuits, such as gates and flip-flops, that is used to process data. It is sometimes referred to as a *data processor*, or simply *processor*.

Data, of course, refers to the binary numbers and words the processor works with. Processing refers to the way the data is manipulated or handled. Types of processing include arithmetic, logic, sorting, translating, editing, counting, and searching. Any action taken on the data is called processing. Processing normally implies that the data is changed in some way or is used to create new data. Data that is not processed as suggested above is dealt with in other ways. Four common ways are storing, retrieving, input, or output. The data is not changed by any of these techniques.

Storing data means putting it in a safe place, like a semiconductor memory or disk drive, where it can be accessed later. *Retrieving*, of course, is the opposite of storing, or going to get the data for reuse.

Input means taking data into the computer to be stored or processed. *Output* means sending the data from the computer to some external device. Input and output (I/O) are ways to transmit data from one place to another.

Another feature of a microcomputer is its *decision-making* capability. During processing, the computer can make decisions and alter its sequence of operation. In other words, the computer can "make up its mind" based on the state of the data or outside conditions.

For example, the computer can tell if a number is less than, greater than, or equal to another. It can choose among alternative courses of actions or say yes or no, true or false, if given enough input facts.

Another major application of computers is *control*. Computers can be used to actuate relays and solenoids or turn lights and motors off and on (see Figure 6.1). In control applications, the computer actually determines when external devices are turned on or off. The computer serves as an electronic clock to time various operations. One example is the embedded single-chip microcontroller inside a microwave oven or a washing machine. Another example is a programmable logic controller (PLC), a special type of microcomputer used

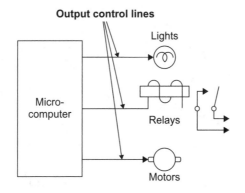

FIGURE 6.1   How a micro controls external devices.

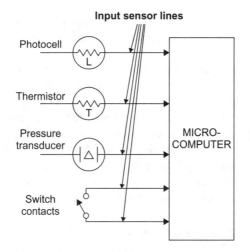

FIGURE 6.2   How a micro monitors external devices.

in industrial applications, that is commonly used to sense, sequence, and time operations in a factory.

The most important thing to remember is that the control is automatic. The micro knows when to perform the various operations that it has been assigned because it has been programmed to do so.

In order to implement certain control applications, the micro must also perform a monitoring function. The micro "looks at" the process or devices being controlled to see what is happening. For example, the computer can monitor switch closures to determine physical state or position, pressure, and many other parameters. Transducers (sensors) convert the monitored physical characteristics, such as temperature or light level, into electrical signals that the computer can understand and respond to (see Figure 6.2). In most microcontrollers or PLCs, the controlled output change takes place only if a specific

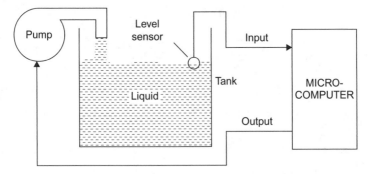

FIGURE 6.3   Controlling liquid level in a tank automatically with a micro.

input condition is sensed or not sensed. For example, if a temperature sensor indicates that the temperature has risen to a specific level, the computer will turn on a cooling fan.

Micros sometimes change the control function in response to one or more of the inputs they are monitoring. For example, if the micro senses that the liquid in a tank exceeds a given level, it can automatically turn off the pump that is filling the tank. If the liquid level goes below a predetermined level, the micro detects this and automatically starts the pump (see Figure 6.3). The key point here is that the computer makes its own decisions based on the input data that it receives. The result is full automation of some process or device.

The micro's ability to monitor and control operation may be used to automate simple processes in a toy or to operate entire factories.

COMPUTER ORGANIZATION AND OPERATION

All microcomputers are made up of four basic sections: memory, control unit, arithmetic-logic unit, and I/O unit. A general block diagram of a microcomputer showing these four sections is given in Figure 6.4. These four sections communicate with one another over multiple parallel electrical conductor data paths called a bus, as shown, to process the data or perform a control function.

Note that the control and arithmetic-logic unit (ALU) are shown together in a common structure. This is the central processing unit (CPU). The CPU is, of course, usually a microprocessor.

The memory is that part of the computer where data and programs are stored. The memory in any computer may contain thousands, millions, or even billions of locations used for storing numbers, words, or other forms of information.

Two primary types of information are stored in computer memory. The first type is the data to be processed. These are the codes, numbers, letters, and other forms of data to be manipulated.

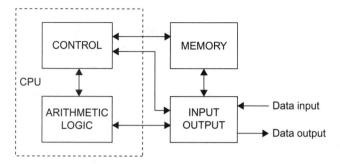

FIGURE 6.4   General block diagram of any digital computer or microcontroller.

The other type of data stored in memory is an instruction. Instructions are special binary numbers or codes that tell the computer what to do. Instructions specify the ways in which the data is to be processed. For instance, there are instructions that cause arithmetic operations to take place or data to be transferred from one place to another.

The instructions listed in a special sequence form a program. A program is a list of instructions that causes data to be processed in a unique way. A program is a step-by-step procedure that solves a problem, performs a control operation, or otherwise manipulates data according to some recipe. Programs are stored in memory along with the data that they process. In some micros there are separate memories for data and instructions. The programs that a computer uses are called software.

Data and instructions can be stored in or retrieved from memory during processing. When data is stored, we say that it is written into memory. When data is retrieved, we say that it is read from memory. A typical write operation transfers data from an analog-to-digital converter and stores it in memory. A common read operation accesses data in memory to be transferred to an LCD screen.

Now let's look at the operation of the CPU. The control unit is that portion of the digital computer responsible for the automatic operation. The control unit sequentially examines the instructions in a program and issues signals to the other sections of the computer that carry out designated operations.

Each instruction is fetched (read) from memory by the control unit, interpreted, and then executed one at a time until the program is completed. This is called the fetch-execute cycle, which is repeated on each instruction until the program runs to completion. The execution of each instruction may call for accessing one or more data words in the memory or storing a data word in memory.

The ALU is the section of the computer that carries out many of the functions that are specified by the instructions. In other words, the ALU actually processes the data. Specifically, the ALU carries out two main types of processing: arithmetic operations (such as addition, subtraction, multiplication, and division) or logic operations (such as AND, OR, complement, or exclusive OR). For example, if an add instruction is stored in memory, the control section will fetch it, interpret it, and send signals to the ALU that cause two numbers to be added.

The ALU also performs data movement operations. It can move data or instruction words from one place to another inside the CPU or it can carry out memory read/write or input/output operations. These are called load and store instructions.

A key part of the CPU associated with the ALU is the registers. Most ALUs have two or more registers. High-powered CPUs may have a group of 16 or more registers, called *general-purpose registers* (GPRs). The registers are used to temporarily store the data being processed by the ALU and the results of the computations. Other registers in the CPU store a number called an *address* where the data or instructions are stored, store the instruction being executed, or act as a stop-off place for data into or out of the CPU.

The control section, ALU, and the registers in all micros are very closely related. They operate together and are always considered as a single unit. As indicated earlier, the combination of the control and the ALU sections is called the CPU (see Figure 6.4). In addition, microprocessors are single-chip CPUs. Besides being called CPUs, microprocessors are sometimes called MPUs or *microprocessing units*. You will also see the expression μP used to refer to a microprocessor. The μ is the Greek letter mu, which means micro; the P means processor.

A microcomputer consists of a microprocessor (the CPU) plus external memory and I/O circuits. Or it could be an embedded controller with everything on one chip. The memory is either SRAM or DRAM for data and programs and EEPROM or flash for the program.

The I/O unit of the computer is the set of logic circuits that permits the CPU and memory to communicate with the outside world. The I/O circuits are referred to as *interfaces*. All data transfers into and out of the computer pass through the I/O interfaces.

The external peripheral devices, or *peripherals*, connected to the I/O unit are electronic or electromechanical units that are used for data entry or data display. Data is most commonly entered into the computer through an input keyboard or a disk drive.

The output data is usually displayed on a video monitor or small LCD screen. "Hard copy" on paper is created by a printer. There are a wide variety of other external input/output devices such as scanners, voice synthesizers, and barcode readers.

In an embedded microcontroller, the peripheral devices include keyboard, ADC, sensors, magnetic stripe readers, barcode scanner, and so on. Outputs go to relays, motors, lights, and liquid crystal displays (LCDs).

OPERATIONAL DETAILS

Registers and the ALU

The main circuit element in a CPU is the register. Usually the data to be processed by the CPU is taken from memory and stored temporarily in a register. Like a memory location, a register usually has a fixed length such as 8, 16, or 32 bits. But unlike a memory location, registers are often used to manipulate data as well as store it. That is, the register can alter or process the data in some way.

Examples of how a register can process data are shown in Figure 6.5. A binary word in a register can be shifted one or more bit positions to the right or left, or the register may be connected as an up/down counter so that it can be incremented (add one to the content) or decremented (subtract one from the content). A register can also be reset or cleared, thereby erasing any data in it and leaving the content zero.

Data transfers and manipulations performed on a register are initiated by individual computer instructions. For example, one instruction may cause the register to be loaded from a memory location. Another instruction may cause the word in the register to be transferred to a memory location. All data transfers are parallel, meaning that all bits are moved simultaneously from the source to the destination.

The main working register in most digital computers is called the *accumulator*. It may also be called the *A register*, *W register*, or something else.

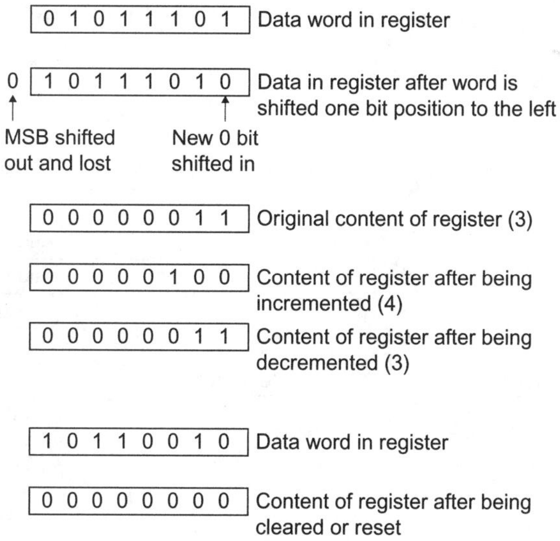

FIGURE 6.5   How data is manipulated in a register. Examples of shifting data, incrementing and decrementing a register, and clearing (resetting) a register to 0.

The accumulator can hold one word whose bit length is equal to that of a memory location. An 8-bit microprocessor has an 8-bit accumulator. The accumulator is part of the ALU.

The accumulator can be loaded from memory or the accumulator content can be stored in any memory location. I/O operations with peripheral units sometimes also take place through the accumulator.

Data to be processed by the arithmetic/logic section is usually held in the accumulator. This data is fed to the ALU. The ALU is a digital logic circuit that can add, subtract, and perform a wide variety of logic operations.

The ALU is capable of processing two inputs, one from the accumulator and one from the memory of another register (see Figure 6.6). Let's assume that you wish to add two binary numbers. To do this, you first load one of the numbers (the *augend*) into the accumulator. This is done with a load-accumulator instruction that takes the number from a memory location and puts it into the accumulator. Next, an ADD instruction is executed. This causes another word (the *addend*) to be taken from memory and placed in register B. It is then added to the content of the accumulator. The sum is usually stored back in the accumulator, replacing the augend originally there.

Some microprocessors have two or more accumulator registers that share a single ALU. Two accumulators provide greater flexibility in the manipulation of data than a single accumulator. Such multiple registers simplify, speed up, and shorten a program to perform a given operation. Some sophisticated CPUs have 4, 8, 16, or even more accumulators. Usually they are referred to as GPRs or a register file. These registers can each use the ALU and act as temporary storage locations for data and the intermediate results of calculations. Instructions are provided to move data from one register to another.

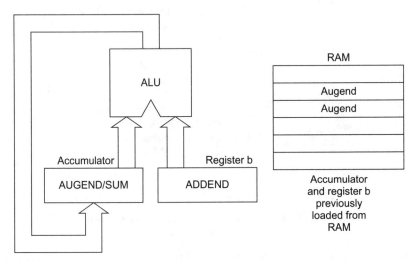

FIGURE 6.6   How ALU adds or otherwise processes two data values.

In most micros, the GPRs share a single ALU. The ALU accepts only two input words and generates a single output word. The two words to be processed by the ALU usually come from any two GPRs or any GPR and a designated memory location. The destination of the ALU output can also be one of the GPRs in some CPUs.

Control Unit

CPUs also have several other registers, including the instruction register; the program counter, also called the instruction counter; and the memory address register, also called the address buffer.

The *instruction register* (IR) is used to store the instruction word. When the CPU fetches an instruction from memory, it is temporarily stored in the IR. The instruction is a binary word or code that defines a specific operation to be performed. The instruction word is also called the *op code* or *operation code*. The CPU decodes the instruction, and then executes it.

The *program counter* (PC) is really a counter and a register. It stores a binary word that is used as the address for accessing the instructions in a program. If a program begins with an instruction stored in memory location 43, the PC is first loaded with the address 43. The address in the PC is applied to the memory, causing the instruction in location 43 to be fetched and executed. After the instruction is executed, the PC is incremented (add 1) to the next address in sequence, or 44. The instructions in a program are stored in sequential memory locations.

The *memory address register* (MAR) or *address buffer* also stores the address that references memory. This register directly drives the address bus and the memory address decoder in RAM or ROM. The MAR gets input from the PC when an instruction is to be accessed (see Figure 6.7). The MAR can also be loaded with an address that is used to access data words stored in memory. To retrieve a data word used in an arithmetic operation, the MAR is loaded with the binary word that points to the location of that word in RAM. This address is often a part of the instruction.

It is important to note that the PC and the MAR (address buffer) have a fixed length of so many bits. And that limits the amount of memory that can be accessed. For example, with a 16-bit address register, the address bus has 16 bits to address RAM and ROM. With 16 bits, a maximum of $2^{16} = 65{,}536$ words can be addressed.

There are usually two other registers, the *flag* and *stack pointer registers*. The flag or *F register* is an 8-bit register whose individual flip-flops are set and reset by the ALU as the various arithmetic and logic operations are carried out. Each flip-flop is called a flag. As an example, there are *zero* (Z) and *carry* (C) *flags*. If the accumulator content is zero after an operation is performed, the Z flag is set indicating this condition. If an arithmetic operation (addition) results in a carry from the MSB of the accumulator, the C flag is set indicating this

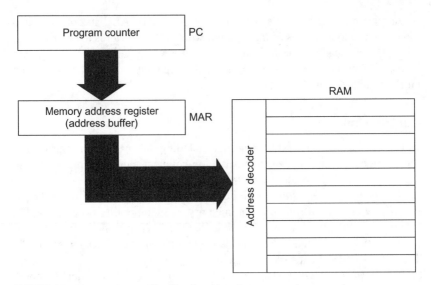

FIGURE 6.7   Program counter identifies the address in memory to be accessed.

condition. These flags can be monitored or tested by the control circuitry to change the sequence of processing.

The stack register is a 16-bit or larger register used to address a selected area of RAM known as the *stack*. This memory is used to store register contents and status information when subroutines and interrupts are used.

Instruction-Word Formats

There are three types of instruction formats used in typical 8-bit microprocessors, illustrated in Figure 6.8. In the single-word format, the instruction is a single 8-bit word. This word is called the *op code*. The op code tells the ALU, the registers, and other elements of the system what to do. In this format, no address is used. The focus of the instruction is implied in the instruction. That is, the data to be processed is already in a location designated by the instruction. Usually the data is in a register. Typical instructions using this format are register-to-register transfer, shift data left (or right), or clear to zero.

The two-word instruction format in Figure 6.8 requires two 8-bit words to define the operation. These two words are stored in sequential memory locations. The first word is the op code. The second word is usually an address that specifies a memory location where the data word to be processed is stored. For example, if the op code calls for an add operation, the address word designates the location in RAM of the number to be added to the contents of the accumulator. The 256 bytes of RAM can be addressed with 1 byte of address.

In some 2-byte instructions, the second byte is not the address. Instead, it is the data itself. This is called an immediate instruction since it is not necessary to address the data that is available immediately within the instruction itself.

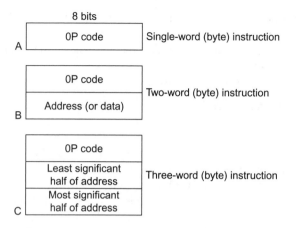

FIGURE 6.8 Three common instruction-word formats with addresses or data.

The three-word instruction format in Figure 6.8 is comprised of an 8-bit op code and two 8-bit address words stored in sequential memory locations. The second and third bytes together form a 16-bit address word that designates the location in RAM of the data to be processed. In the 3-byte instruction format, the first byte is the op code, the second byte is the least significant half of the address, while the third byte is the most significant part of the address. Different formats are used by other micros.

To access a word in RAM, the instruction address word must be stored in the MAR. This happens during the instruction fetch operation. When an instruction is fetched from memory, the op code is stored in the instruction register while the address is stored in the MAR. The instruction is then executed. The MAR usually gets its input from the PC. Once an instruction is fetched and executed, the PC is incremented.

The PC may be incremented once, twice, or three times, depending on the length of the instruction just executed. If a 2-byte instruction is executed, the PC is incremented twice so that the PC points to the address of the next instruction op code.

Program Execution Example

Now let's illustrate how a typical CPU executes a simple program. Here are a few common instructions similar to those found in any micro.

Move A to B (MOV B,A): This is a 1-byte instruction that causes the content of the accumulator (A) to be moved to register B. The content of the accumulator is not erased.

Load Accumulator (LDA): This is a 3-byte instruction that takes the content of the address specified by the second and third bytes and loads it into the accumulator.

Add Immediate (ADI): This is a 2-byte instruction that takes the content of the second byte of the instruction and adds it to the content of the accumulator. The sum is stored in the accumulator, replacing the number that was there previously.

Halt (HLT): Halt is a 1-byte instruction that stops processing.

Note that we refer to each instruction by a three-letter abbreviation called a *mnemonic*.

Figure 6.9 shows a complete CPU and the RAM containing a simple program. The first instruction in the program is the LDA stored in location 0000.

To begin the execution, the address of the first instruction is loaded into the PC that points to the location of the instructions in RAM.

Next, the content of the PC is transferred to the MAR that addresses RAM. The LDA instruction op code is fetched and stored in the instruction register. The address bytes of the LDA instruction in locations 0001 and 0002 are then transferred to the memory address register. The address of the data to be loaded into the accumulator in bytes 2 and 3 is 0008. The desired data word in location 0008 is retrieved and loaded into the accumulator. This is the augend (23).

After the LDA instruction is executed, the number 23 is stored in the accumulator. Next, the PC is incremented three times. The ADI instruction is then fetched and stored in the instruction register. The number to be added is stored in location 0004.

This is an ADD immediate instruction where the second byte contains the data, in this case the addend (56). The number in location 0004H is 56, which is added to 23 in the accumulator, creating the sum 79. The mnemonic of the next instruction to be executed is MOV B,A in location 0005.

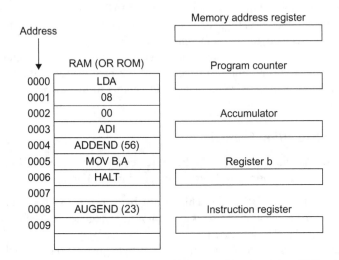

FIGURE 6.9 Example of program stored in RAM for execution using various CPU registers.

The sum in the accumulator is then moved to the B register by the MOV B,A instruction. The content of the accumulator is not changed when data is transferred to the B register.

The PC is incremented once since the HLT is a 1-byte instruction.

As you can see by this step-by-step analysis of the execution of a computer program, there is nothing mysterious about its operation. The CPU simply fetches and then executes the sequentially stored instructions at very high speed until the operation is complete. This process is sometimes referred to as the *stored program concept*.

Microcomputer Buses

Data transfers inside a microprocessor take place in parallel. This means that all bits in a word are transferred simultaneously from one place to another. It takes only a few nanoseconds (one billionth of a second) for all data bits in one register to be moved to another register.

The parallel data transfers take place over a data bus. A bus is simply multiple parallel electrical connections from a source to a destination. A number of these buses are contained within the CPU and are known as internal buses.

Microprocessors usually have three major buses—a *data bus*, an *address bus*, and a *control bus*. These are made available to external circuits in some micros. A typical 8-bit CPU has an 8-bit data bus and a 16-bit address bus. The data bus sends data to and from the CPU, RAM, ROM, and I/O sections. All data transfers between the CPU and memory or I/O sections take place over the data bus. The address bus drives all of the memory and I/O devices.

When an instruction is fetched from RAM or ROM, it is transferred over the data bus from the memory into the instruction register. Any data word retrieved from memory or a peripheral device via the input section also passes over the data bus into the accumulator or GPR.

When a store instruction is executed, the word in the accumulator is transmitted over the data bus into RAM. Data can also be transferred from the accumulator over the data bus to an external device such as an LCD. Or, data from an input device such as a keyboard passes over the data bus and is placed into the accumulator. The important point here is that data can move in either direction over the data bus. We say that it is a *bidirectional bus*.

Another key point is that the data bus can be connected to only one data source at a time. Data can originate only at a single source, but it can be sent to one or more destinations. To accomplish this, circuits called *bus multiplexers*, or *three-state line-driver circuits*, are used to connect or disconnect the various data sources to or from the bus. Figure 6.10 illustrates this concept.

The address bus is a unidirectional bus. It transfers an address from the CPU to all external circuits (memory and I/O). Address words are produced in the CPU. The PC generates the address that points to the instruction to be fetched. The content of the PC is transferred over a parallel 16-bit internal

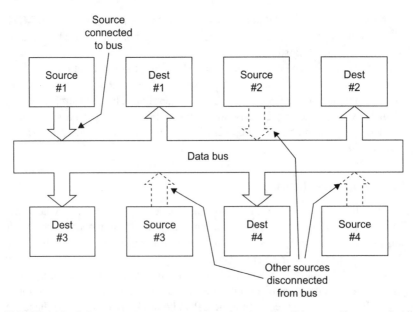

FIGURE 6.10 Only one source may transmit at a time on a bus. Others are disconnected. All destinations can receive the transmitted data. Dest=destination.

address bus to the memory address register or address buffer. The output of the MAR or address buffer is the address bus.

The control bus consists of numerous signals generated by the microprocessor and used to initiate various memory or I/O operations. The control bus also contains input lines from external circuits that tell the CPU what to do and when. The number of control signals varies from CPU to CPU.

POPULAR MICROS

The first practical microprocessors and microcomputers became available in the mid-1970s. Over the years lots of different models have been developed. As semiconductor processing technology has improved over the years, it has become easier and cheaper to put more circuits on a chip plus the memory that earlier was always external. And processing speeds have multiplied many times over. Today's embedded controllers are indeed complete computers on a chip and they run at very high clock rates. And their cost is low. In high-volume manufacturing, an embedded controller can cost less than $1.

Most of the initial processors were 8-bit devices. These are still popular today. Larger, faster processors are available for not much additional cost. Sixteen- and 32-bit processors are commonplace. Even 64-bit devices are available if needed. Many are individual chips, while others are available only as cores on a larger systems chip. Following is a summary of a few of the most widely used processors.

8- and 16-Bit Microcontrollers

8051

The 8051 is one of the oldest but still widely used. It was originally developed by Intel in the late 1970s and 1980s. It is still around and available from a half dozen or more companies. The more recent versions are faster and have more memory and I/O features than the original. Original clock speeds were only 8 to 12 MHz, but today you can get models with a clock rate up to 100 MHz.

A general block diagram of the 8051 is given in Figure 6.11. The 8051 features an internal RAM of 128 or 256 bytes but can handle 64 or 128 KB of RAM externally. The internal ROM can be had from 512 bytes to 128 KB. Some models also feature flash memory. As for I/O ports, the figure shows four 8-bit parallel I/O ports. You can get a standard serial port called a UART, and more specialized I/O ports such as SPI and I²C. See the Interfaces sidebar in this chapter. Many models also come with an internal 10- or 12-bit ADC.

A few other items in this diagram that were not previously discussed are the clock, interrupts, and timers. The *clock* is an oscillator that controls all operations in the micro. Its frequency is usually set by an external crystal that maintains a precise and stable frequency. The higher the frequency, the faster the operations. Fast is good.

An *interrupt* is, as its name implies, an external signal that can interrupt the CPU when an important event occurs. It can signal the CPU of the event and even cause the CPU to stop executing the current program and jump off to another program to deal with the interrupt.

Timers are usually just one or more up/down counters than can be programmed with various count values. These are used to implement count functions in programs and to provide timing operations for I/O operations. Most controllers have one or more timers.

68HC11

Another widely used 8-bit controller is the Freescale 68HC11 and its various versions. There are dozens of offshoots and variations. It has two 8-bit accumulators and one 16-bit accumulator. Total memory space is 64 KB with 128, 256, or 512 bytes of RAM, and up to 48 KB of ROM program memory. Different versions have all the interfaces, both parallel and serial, and some models have a built-in 8-bit ADC. A 16-bit version called the 68HC12 is also part of the product mix.

PIC Processors

Microchip Technology makes one of the most widely used 8-bit controllers. Called PIC processors, they are extremely simple and cheap. There are hundreds of versions with different memory sizes and configurations and various mixes of interfaces. One of the simpler versions is the 16C54. It has a single

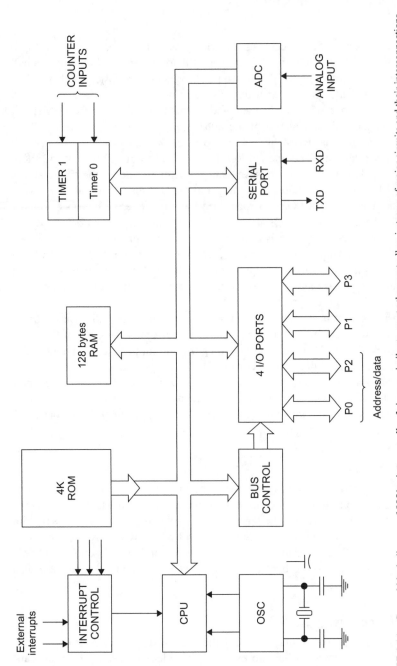

FIGURE 6.11   General block diagram of 8051 microcontroller. It is very similar to most other controllers in terms of major circuits and their interconnections.

working register, W, that is the accumulator. It uses a bank of 24 to 73 registers for RAM. ROM for program storage ranges from 512 bytes to 2 KB. I/O occurs via 4- or 8-bit parallel ports. Larger versions have more RAM and more ROM, higher clock speeds, a built-in ADC, and special network interfaces like CAN. Several 16-bit versions are also available.

MPS430

A typical 16-bit microcontroller is the Texas Instruments MSP430. It has 16 GPRs and a clock speed of 16 MHz. On-chip RAM is SRAM, and can be had in versions with 128 bytes to 10 KB. ROM for the program is flash from 1 KB to 120 KB. It has an ADC on board, a DAC, timers, and all the common serial interfaces. It is a popular chip because its low power consumption makes it ideal for portable and battery-powered devices.

32- and 64-Bit Processors

As it turns out, it is almost as cheap and easy to make a 32-bit processor as it is to make an 8-bit device. For that reason, for a few cents more, designers can make their products faster and more powerful by selecting a 32-bit device. With this many bits being processed at the same time, it is possible to move more data faster, compute with increased resolution, and to perform much more complex operations faster and easier. Thirty-two-bit processors can also address much more memory. A few years ago, 32-bit processors were reserved for only the largest or most critical of needs. Today, the 32-bit processor is almost as widely used as the 8-bit processor.

Three devices account for most applications. They are made by ARM Holdings, Freescale, and MIPS. ARM is the name of a British company that invented a super-simple, fast, and flexible 32-bit processor that does everything well at very low cost. It is referred to as a reduced instruction set computer (RISC) processor. It has a few simple instructions but is extremely fast. ARM does not actually make chips, but instead licenses the design to others. There are several vendors of ARM processors but most of them are actually cores that are integrated on-chip with other circuits to form a system on a chip (SoC). With over 75% of the market share, you are sure to encounter an ARM device sooner or later. Chances are you own at least one inside a product such as a cell phone, which is where most are used.

Freescale's PowerPC is another popular 32-bit processor. It was originally used in Apple Macintosh computers but no more. IBM used it widely and still does in some products. Many embedded versions have since been made. Its primary usage is in communications, networking, automotive, and industrial applications.

For really super processing power, the MIPS processors are a popular choice for certain embedded applications. These devices come in 32-bit as well as 64-bit models, in case you really have a hot application for which nothing else will do. The MIPS processors are great at advanced math and have excellent resolution.

Microcontroller Interfaces

An *interface* is the circuitry that connects an embedded controller to the outside world. Initially, all I/O in early micros was just the parallel data and address buses extended to accommodate additional I/O circuitry. Today, much of the I/O interface circuits are put right on the same chip with the rest of the micro. Following are the most widely used interfaces you are likely to encounter. All of these are serial interfaces. Parallel I/O is still used, but today the preference is serial I/O because of fewer interconnecting lines.

RS-232/UART—This universal asynchronous receiver-transmitter is actually the physical circuitry implementing a popular serial interface standard referred to as RS-232. It was originally developed for connecting computers to teletype machines and early video terminals. It is still used today, especially for connecting to industrial equipment for monitoring or control.

The RS-232 interface sends and receives data in bytes (8 bits) plus two extra bits, called Start and Stop bits, that tell the receiving device when the data begins and ends (see Figure 6.12). The logic levels are −3 to −25 volts for a binary 1 and +3 to +25 volts for a binary 0. Typical voltage levels are in the 5- to 12-volt range. The data rate is slow, typically from 300 bps to 115.2 kbps. Standard data rates are used, so no clock signal is needed. Common speeds are 2400, 9600, and 19,200 bps. The physical communications medium is a two-wire (signal plus ground) cable that can be up to 50 feet long. Special 9- and 24-pin connectors are part of the standard. Many PCs and laptops still have a 9-pin RS-232 port.

I²C—Shorthand for inter-integrated circuit or I squared C, this is a low-speed serial bus that is used for sending data from one chip to another on the same PC board or over short cables between two pieces of equipment. The data rate is typically in the 10 kbps to 400 kbps range, but a faster version (3.4 Mbps) is

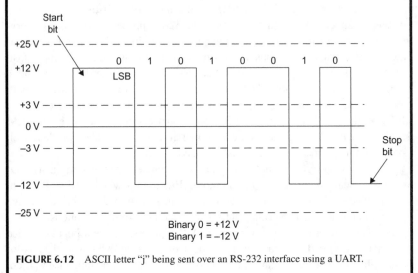

Binary 0 = +12 V
Binary 1 = −12 V

FIGURE 6.12 ASCII letter "j" being sent over an RS-232 interface using a UART.

(Continued)

Microcontroller Interfaces (Continued)

available. Speeds rarely exceed 100 kbps. Only three wires are needed, data, a clock signal that controls the data transfer, and a ground.

SPI—This is the serial peripheral interface. It uses four wires, data in, data out, clock, and a select signal that determines whether a device is a master or slave and either sending or receiving data. A fifth connection is ground. It is much faster than the I^{2}C interface, with speeds up to 20 Mbps.

CAN—This is controller area network serial bus in an interface used to link micros together in small networks. It is used in automotive and industrial applications. The usual cable is twisted pair and a ground. Speeds range from 10 kbps to 1 Mbps with 20 kbps being typical. The data is sent in specifically formatted frames as determined by a standard protocol. It uses start and stop bits like the RS-232 and some similar control codes defined in ASCII. It also includes error detection and correction capability, so it is very reliable.

USB—The universal serial bus is the most common PC interface for peripherals including keyboard, mouse, printer, flash drive, external disk, and so on. It is also used in some embedded micros. The interface uses two wires but also includes +5 volts DC and ground. Version 2.0 provides speeds to 480 Mbps over cables no longer than 16 feet. A new higher-speed version 3.0 is also available with speeds up to 5 Gbps. Data is sent in bytes in packets according to a unique protocol.

RS-485—This is an interface widely used in industrial applications. It uses twisted pair cable that can be up to 4000 feet long. The cable is actually a bus to which many devices can be attached. Data rates can be up to 10 Mbps for shorter lengths. Various protocols may be used.

A WORD ABOUT SOFTWARE

The microcontroller can perform almost any digital function from the simplest logic operation to a complex mathematical calculation. The only problem is that you have to tell it to do what you want. And you do that by programming. A program is the sequence of instructions that detail the order in which you do the various processing steps that lead to your application. The controller chip itself is a dumb piece of hardware waiting for its program. What this means is that today, designing an electronic product is as much about writing programs as it is building electronic hardware.

Software is the name given to all the programs that a processor uses for developing and executing a program. The program that implements your application is the applications program. Other types of software used with the processor follow:

Languages—A programming language is a set of rules or syntax that defines how you tell the processor what to do. A language is the tool that you actually use to create your application program. It is a kind of shorthand code that lets you express the operations in simple terms.

Editor—An editor is like a word processing program that lets you type in your program. You then save it as a file that the compiler or assembler uses to create machine code.

Compiler—A compiler is the program that translates your program code into binary code that can be stored in the memory and used by the processor. Processors only understand binary code but humans are not good with such code. So, you write your program in a language you can understand and the compiler converts it to the code that the processor understands.

Operating system—For large complex programs or for applications that require the processor to execute multiple programs at different times, you need a piece of software that manages the multiple programs and operations. The operating system (OS) oversees things like memory usage and access, timing, and I/O operations. Not all embedded controllers need an OS, just the more powerful ones with multiple functions. For example, most cell phones have an OS.

Development environment—This is a complete package of software tools that facilitates your software development. It consists of languages and compilers, as well as programs for testing and debugging your code.

Subroutines—These are short sections of code that perform a specific operation. It may be a math operation, some type of I/O process, or other function that is performed multiple times. You only write the subroutine once, and then access it when you need to do that operation. Many development systems have libraries of subroutines that you can tap so you do not have to program common functions yourself. Subroutines shorten your development time and speed execution.

Programming Languages

There are dozens of languages used from programming embedded controllers. The more popular ones are assembly language, C, and BASIC.

Assembly Language

Assembly language uses shorthand mnemonics such as three- or four-letter designations for the instructions that the processor can execute. Instead of programming with binary code, you use the mnemonics. In addition, you use names that you make up to identify memory locations, I/O ports, and other things normally referenced by a binary code. Following is what a simple assembly program looks like:

```
ORG $0000
LDAA FIRST
MUL SECOND
STA PRODUCT
OUT PORT 1
END
```

This programs loads one number from a memory location given the name FIRST, multiplies it by another number in memory location named SECOND. It then stores the result in another memory location called PRODUCT. It next outputs the value to an I/O port called PORT 1. The ORG $0000 simply says to start the program whose first instruction is in memory location 0000, a hex address.

The software that converts the assembly language program to binary code is the *assembler*. The assembler gives actual binary assignments to the memory locations named as well as the instruction op codes.

C Language

The C language is referred to as a higher-level language in that the various syntax and terminology of the language usually translates into multiple binary machine instructions. The C language lets you write the program in a more conversational or casual format using English instead of cryptic mnemonics and other terms. The language is a bit of a challenge to learn, but after that programming goes much faster. A compiler converts the C-language program code into binary code.

A simple C program may appear like this:

```
int x, y, z
x = 38
y = 57
z = 57 - 38
printf z
```

The first line identifies three integer values x, y, and z; 38 is assigned to x, and 57 to y. Then the difference between the two is designated z. Finally, the printf line accesses a subroutine that causes z to be printed.

BASIC Language

Some embedded controllers use a simplified version of BASIC. BASIC was invented before personal computers and initially became the main language of personal computer programming. Microsoft still sells its Visual BASIC software for program development. BASIC is very simple higher-level language that is easy to learn and use. The program created is then translated into binary code by a compiler. Alternately, the controller runs a program called an *interpreter* that resides in memory and executes the program a command at a time.

The following program is written in PBASIC, a simple version of BASIC used with an embedded controller called the BASIC Stamp made by Parallax. It uses a PIC microprocessor and stores a program in flash memory. The program causes an LED to flash off and on.

```
Flash
Output 7
High 7
Pause 500
Low 7
Pause 500
Goto Flash
```

The name of this program is Flash. The first line tells the controller to make pin 7 the output port. The next line tells pin 7 to go high or to binary 1. The next command says pause for 500 milliseconds. Next, pin 7 is made to go low or to binary 0 via the Low command. Pause makes it stay there for another 500 milliseconds. Finally, the Goto command causes the program to look back and repeat itself again and again. The result is an LED that turns off and on at a 1000-millisecond (1-second) rate.

DIGITAL SIGNAL PROCESSING

Digital signal processing (DSP) is the technique of programming a microcomputer to perform operations normally carried out by analog or linear circuits. The method is illustrated in Figure 6.13. Assume you want to process an analog signal by filtering. You could use a hardware filter made up of inductors and capacitors or resistors and capacitors. Another way to do it is with DSP. The analog signal to be filtered is first digitized with an ADC and the resulting data stored in a RAM. This data is then acted upon by a processor programmed to perform a filter function. Special mathematical algorithms have been developed to do this. The resulting calculations generate new data that is also stored in the RAM. This new data is then sent to a DAC where it is converted back to an analog signal. The result is that the signal is filtered as if were processed by an electronic circuit of capacitors and inductors. The processing was mathematical.

There are many other operations that can be performed by DSP. Filtering is the main application, but you can also do things like modulation and demodulation as needed in wireless applications. It will perform other things such as data compression, equalization, tone control, and even frequency analysis.

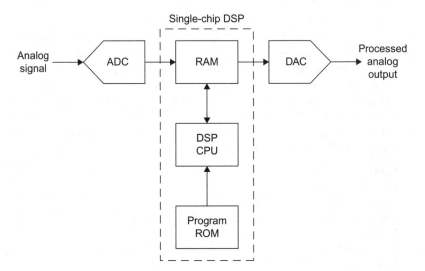

FIGURE 6.13   Basic make-up of digital signal processor and its ADC and DAC I/O.

Using a program called the fast Fourier transform (FFT), the DSP can analyze a signal and tell you exactly what frequency components are in it. It identifies the harmonic frequencies and their various amplitudes and phases just as in a frequency-domain plot.

The big question is why use a complicated technique such as DSP when you can do it with simple analog/linear hardware? Probably the best answer is "just because we can." With cheap single-chip processors, DSP is often no more expensive than an equivalent analog circuit. Since everything else is digital these days, it makes some sense to do as many analog functions digitally as well. But the best reason is that DSP provides superior results over analog methods. Filtering is substantially improved with better selectivity that was not possible with analog filters.

While DSP can be programmed on any processor or controller, it is best done on special DSP chips designed for the purpose. They use special architectures and processing schemes to speed up and improve the processing over standard conventional processors. A good example is that DSP chips implement a multiply and add or accumulate function, called MAC, that is common to most DSP processes. The MAC has to be programmed in a standard processor, but in a DSP chip it is implemented with special hardware that speeds up the process and simplifies programming. Some more conventional processors also incorporate DSP instructions to facilitate DSP operations.

Now that DSP chips are cheap despite their complexity and speed, DSP is widely used. It is found in most electronic products today, such as TV sets, cell phones, MP3 players, cable modems, and many other common products.

Project 6.1

Use a Microcontroller
One of the fastest ways to get familiar with a microcontroller is to use the BASIC Stamp made by Parallax. It comes as a complete unit with a microchip technology PIC processor and I/O circuits. One is a USB port to a personal computer. The software provided lets you program the controller in PBASIC. You write the program on a personal computer and download it to the EEPROM in the PIC. Then you run the program. It is ideal as a way to monitor and control lots of simple devices. And it is a fast and easy way to see firsthand what a controller does.

Go to the Parallax website at www.parallax.com and find the BASIC Stamp info and just order one. There are various models, from the ultra simple to the complex, and lots of accessories to help you implement various applications.

Project 6.2

Learn More about DSP
Go to the Texas Instruments website at www.ti.com. TI is the leader in DSP chips and has lots of useful information for your learning purposes.

Chapter 7

Radio/Wireless

The Invisible Cables of Modern Electronics

In this Chapter:
- The oldest electronic application.
- What are radio waves?
- The electromagnetic frequency spectrum.
- The hardware of radio: transmitters, receivers, and antennas.
- Two-way radio applications.
- Satellites.
- Radio telescopes.

INTRODUCTION

First they called it wireless. Then they called it radio. Today, the term *wireless* is trendy again. Whatever you call it, radio is communication over a distance without any physical connection. Radio is cool because it is completely silent and invisible. It can occur over a distance of only a few feet, all the way around the world, or between planets and spacecraft. Radio, despite the fact that it has been around for over a century, and the fact that we take it for granted on a grand scale, is truly black magic. When you finish reading this chapter, you will know how this magic works.

RADIO COMMUNICATION SYSTEMS

Radio is an electronic communications system. All electronic communications systems have the basic form shown in Figure 7.1, whether it is a wired or wireless system. The information to be communicated, usually voice, is first converted into an electrical signal by a microphone. That signal is then processed by the circuits in a transmitter (TX), and then sent via a communications medium to a receiver (RX) that processes the signal and converts it back into voice by a speaker or headphones. In the telephone system, the communications medium is a cable made of copper wire. In a cable TV system, the medium is a coaxial cable, twisted-pair cable, or fiber optic cable. But in radio, the communications channel is free space. The radio signal travels from one place to another almost instantaneously and it passes right through

doi: 10.1016/B978-1-85617-700-9.00007-2

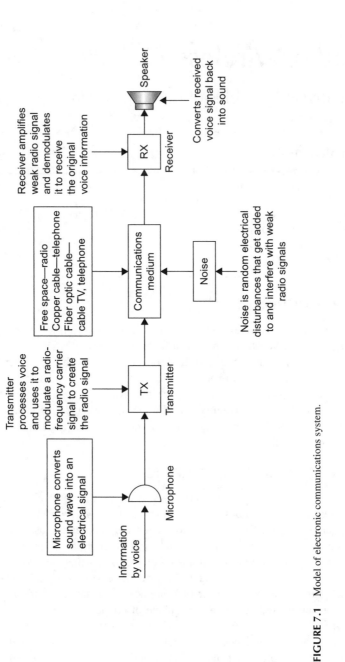

FIGURE 7.1 Model of electronic communications system.

most obstructions like walls, buildings, smog, trees, and even people. Note that noise is added to the signal as it passes through the channel or medium.

Noise

Noise is any random electrical voltage that interferes with the weak radio signal. Noise comes from the atmosphere, space, human-made objects, and electronic components. Atmospheric noise is lightning. Space noise comprises random signals from the Sun and remote stars. Human-made noise comes from electrical power generation, distribution, and switching. It also comes from fluorescent lights, motors, auto ignitions, and many other sources. The worst noise comes from the electronic components that actually process the weak radio signal. Resistors, transistors, and integrated circuits generate random noise because of thermal agitation. Heat produces random movement of the electrons in the components, and that produces a noise voltage which is added to the radio signal. Even though noise voltages are very small (microvolts or nanovolts), they are still large enough to interfere with and in some cases obliterate the very small received radio signal. In most radio systems, it is noise that limits intelligibility, transmitted data speed, and transmitted distances.

THREE BASIC WAYS OF COMMUNICATIONS

Radio communications systems work in one of three basic ways, simplex, half duplex, and full duplex (see Figure 7.2).

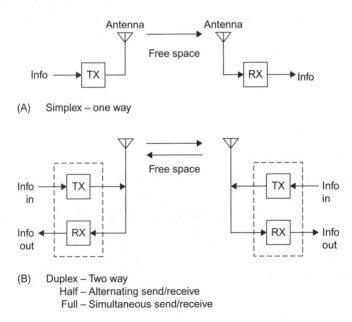

(A)   Simplex – one way

(B)   Duplex – Two way
        Half – Alternating send/receive
        Full – Simultaneous send/receive

FIGURE 7.2   Ways of communicating.

Simplex

Simplex is one-way communications. Examples are radio broadcasting, paging, and remote control. A radio telescope is also one-way: there is a transmitter at one end and a receiver at the other.

Half Duplex

Duplex means two-way, and most radio communications are really two-way affairs. There is a transmitter and a receiver at both ends of the communications system. Both parties can send and receive. And there are two types of duplex systems, half duplex and full duplex.

In half-duplex systems, only one party can transmit at a time. Those communicating with one another take turns talking. This is the kind of radio where you have to say "over" or "come back" after you finish speaking to let the other party know that it is time for her or him to talk. CB radio is like this. Most two-way communications services (police, fire, taxi, aircraft, marine, etc.) use this method because it is cheap and simple. Fax machines are half duplex. Most computer modems are half duplex.

Full Duplex

Full duplex is more complex and expensive. This is two-way radio where both parties can both send and receive simultaneously. Probably the best example of this is a cordless or cellular telephone. It is great to be able to talk and listen at the same time.

WHAT IS A RADIO WAVE ANYWAY?

Any discussion of radio always gets around to this basic question. How do you describe something that you cannot see, feel, or hear? You imagine it.

Electric and Magnetic Fields

A radio wave is an invisible force field that is made up of both a magnetic field and an electric field. In combination, we call them an *electromagnetic field*. We can't see or feel them, but we can experience their effects. In Chapter 2, electric and magnetic fields are described briefly. To review, a magnetic field is a force generated by a magnet, either a permanent magnet or electromagnet. Current flow in a wire or other conductor produces the magnetic field. Invisible magnetic flux lines emanate from the magnet and may influence objects in or near them.

An electric field derives from a voltage, or what we call a *potential*, across two conductors. The attraction of the positive and negative charges across an open space introduces another kind of invisible field that can also influence external objects.

Now if you combine these two types of fields into a self-supporting entity, you would have an electromagnetic field or *radio wave*. We create a radio signal with the circuits in a transmitter, but it is the magic of a device called an *antenna* that produces the combined electric and magnetic field that we refer to as a radio wave.

In the 19th century, a British scientist named James Clerk Maxwell figured out that if you generate a changing electric field, it would, in turn, generate a magnetic field. If the magnetic field is changing and moving forward, it will generate an electric field. The two fields support or regenerate one another as they change and move outward from the antenna that produces them. Maxwell expressed his theories in his famous mathematical equations that form the foundation of radio and light transmission.

Trying to draw a picture of a radio wave is somewhat like trying to draw a picture of the wind. Nevertheless, Figure 7.3A is an attempt. Try to think in three dimensions. Current flow in the antenna wire creates the magnetic field lines. The voltage across the two segments of the antenna produces the electric field lines. The picture is somewhat simplified if we just show the field lines as in Figure 7.3B. The electric and magnetic field lines are always at right angles (90 degrees) to one another. The radio wave moves in a direction at a right angle to both field lines, in this case either into or out of the page. As it gets some distance from the antenna, it becomes self-supporting. That is, the magnetic field sustains the generation of the electric field and vice versa.

Polarization

Polarization of a radio wave means how it is oriented with respect to the surface of the Earth. If the electric field is vertical to the surface of the Earth, the wave is vertically polarized. If the electric field is horizontal to the Earth, the wave is horizontally polarized. It is the orientation of the antenna generating the radio signal that determines the polarization. A vertical antenna produces vertical polarization, and a horizontal antenna produces horizontal polarization. For the best reception, the transmitting and receiving antennas should both be either vertical or horizontal.

There is also circular polarization. Some antennas make the signal rotate clockwise or counterclockwise as the signal travels from transmitter to receiver. Clockwise rotation, as viewed from the transmitter, is called right-hand circular polarization (RHCP). Counterclockwise rotation is called left-hand circular polarization (LHCP).

Signal Speed

Radio waves travel at the speed of light in space. That is 300,000,000 meters per second or 186,410 miles per second. So while radio transmission is not

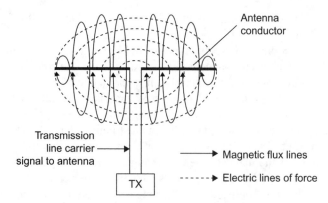

(A) Antenna produces both electric and magnetic fields

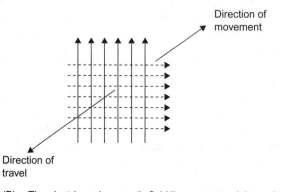

(B) The electric and magnetic field lines are at a right angle
to one another and to the direction of radiation

FIGURE 7.3 How an antenna creates a radio wave.

instantaneous, it is pretty fast. Just as an example, to transmit a signal to the moon that is 250,000 miles away takes $250,000/186,400 = 1.34$ seconds.

Signal Strength

As the radio wave gets farther and farther from the antenna, it gets weaker and weaker. The power in a radio wave, in fact, decreases by the square of the distance from the antenna. For example, doubling the distance decreases the received power by two squared (2^2) or four. Even when the transmitter puts many hundreds or thousands of watts of power into the antenna, the signal received at a distance is many times smaller. Typical radio signals are usually expressed in terms of microwatts (1-millionth of a watt) or nanowatts (1-billionth of a watt).

As the radio wave encounters a receiving antenna, the magnetic and electric fields induce a voltage into the antenna. This tiny voltage is then amplified and translated back into the original information signal by the receiver.

ELECTROMAGNETIC FREQUENCY SPECTRUM

The radio signal applied to the antenna by the transmitter is a sine wave. Remember that all sine waves have both an amplitude and a frequency. Frequency is stated in terms of cycles per second or hertz (Hz). Radio transmissions can take place over a very wide frequency range. We call this range the *electromagnetic frequency spectrum*. It extends from approximately 10 kHz to 300 MHz. Figure 7.4 shows a diagram designating the major divisions of the electromagnetic spectrum.

The electromagnetic spectrum is in free space, not only here on Earth but throughout the universe. The spectrum space may be used by anyone and, in fact, on Earth we all share it. The spectrum is divided up and used for many different radio services. It is also partitioned such that various frequency ranges are assigned to or used by different countries for their various radio services. In many cases, portions of the frequency spectrum are shared concurrently among several countries. This allocation of the frequency spectrum is necessary to prevent radio stations from interfering with one another.

In the United States, the frequency spectrum is regulated by the Federal Communications Commission (FCC). Congress established the Communications Act of 1934, which has subsequently been amended to permit the FCC to regulate all forms of radio and wired (telephone, cable TV, etc.) communications services. The National Telecommunications Information Administration (NTIA) manages the spectrum for the government and the military. Throughout the world, the frequency spectrum is allocated and managed by a worldwide organization known as the International Telecommunications Union (ITU).

Major Segments of the Frequency Spectrum

30 kHz to 300 kHz—Low frequencies (LF)
300 kHz to 3 MHz—Medium frequencies (MF)
3 MHz to 30 MHz—High frequencies (HF) or shortwaves
30 MHz to 300 MHz—Very-high frequencies (VHF)
300 MHz to 3 GHz—Ultra-high frequencies (UHF)
3 GHz to 30 GHz—Super-high frequencies (SHF)
30 GHz to 300 GHz—Extremely high frequencies (EHF)

The various areas of the frequency spectrum are assigned to different types of communications services. For example, in the low-frequency (LF) range, most communications services are navigational systems used by the Coast Guard and military. The AM broadcast band (530 kHz to 1710 kHz) is in the

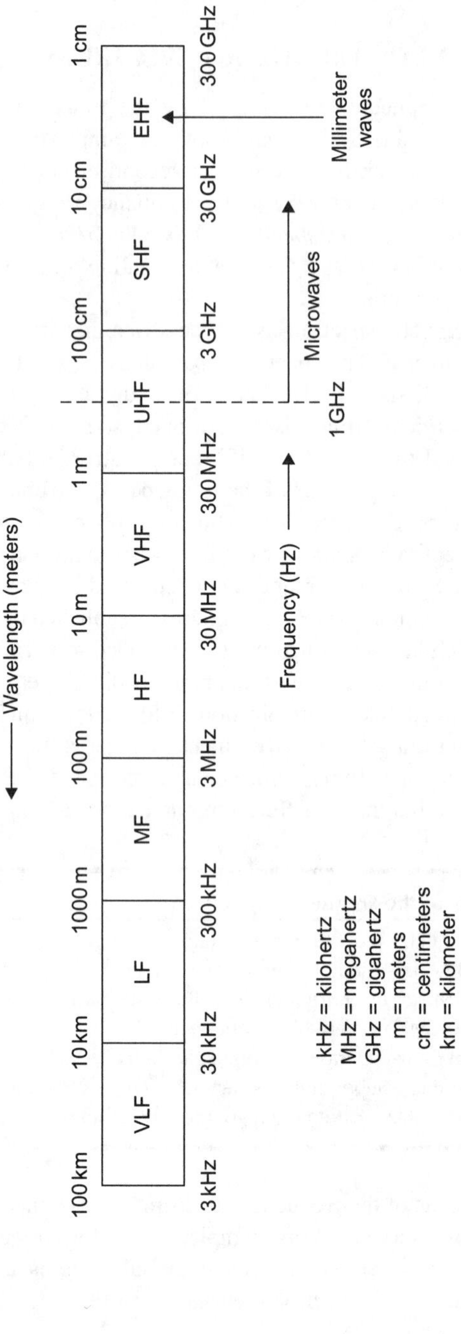

FIGURE 7.4 Electromagnetic frequency spectrum.

medium-frequency (MF) range. Shortwave (SW) radio broadcasts, CB radio, amateur radio, and a variety of other communications services are in the 3 to 30 MHz or high-frequency (HF) range.

One of the most widely used areas of the spectrum is the VHF range from 30 to 300 MHz. Virtually all two-way radio communications for marine, aircraft, mobile, and other services are in this range. FM broadcasting (88 to 108 MHz) and TV channels 2 through 7 are also in this region.

The UHF frequency range, from 300 MHz to 3000 MHz (3 GHz), also contains a wide variety of communications services. It includes two-way mobile radio, TV, and cellular telephone.

Frequencies above 1 GHz are designated as microwaves. In the microwave region between 1 and 3 GHz are satellite and navigational systems. For example, the widely used global positioning system (GPS) operates in this range. This area is also occupied by the newer cellular telephone service and wireless networks such as Wi-Fi and Bluetooth.

The frequencies from 30 GHz to 300 GHz are called millimeter waves because their wavelength is in millimeters. These frequencies are typically used only by satellite and radar systems. Other services are point-to-point wireless data services and cellular backhaul. Some short-range wireless devices have also been defined for the 60-GHz range. These will be used in consumer applications as well as in faster wireless networks.

Critical Issue: Spectrum Space

There is only so much radio spectrum space and we must all share it. It is a true natural resource. That's why spectrum space is regulated by governments. But we have just about used it all up. Virtually every hertz of space is spoken for. Diverse radio services compete for space. In some cases, an old service must die before a new one can begin. Sometimes space is stolen from one service to give to another, such as in the mid-1980s when the upper UHF TV channels were sacrificed to make room for cellular telephones. Lately, the U.S. government has been selling space to the highest bidders for new cell phone and other wireless services. During the recent switchover to digital TV in the United States, a huge chunk of the spectrum was freed up (108 MHz, from 698 to 806 MHz) as TV stations abandoned this region. Known as the 700-MHz band, it was auctioned off by the government for $19.6 billion for more cell phone spectrum as well as broadband wireless services and cell phone TV.

The trend has been to push higher and higher in the frequency range to find more space. During the past several decades, the microwave region (above 1 GHz) has been developed. This region, in the UHF, SHF, and EHF ranges, was previously used only for radar and satellites. But now it is being used and developed for digital cell phones, navigation, high-speed networking, broadband (Internet) wireless, and digital TV.

Beyond the millimeter wave range above 300 MHz is the optical spectrum. That is, above 400 GHz, the electromagnetic radiation becomes light, namely infrared, visible light, and ultraviolet light. But that's another story altogether.

Wavelength

While we normally use frequency to refer to location of a radio signal in the spectrum, sometimes you will hear about frequencies or ranges of frequencies designated by wavelength.

Wavelength is the actual physical length of a radio wave. It is the amount of space the wave occupies. Specifically, wavelength is the length of one cycle of the sinusoidal variation of the electromagnetic fields. If we could see the electric and magnetic fields, we could measure the distance between the maximum intensity points in meters (1 meter = 3.28 feet or 39.37 inches). This is the wavelength. Wavelength is also the distance that the radio wave travels in the time of one cycle.

You can determine the wavelength of a radio signal with the following simple formula:

$\lambda = 300{,}000{,}000/f$

Wavelength is represented by the Greek letter lambda (λ) and frequency in hertz is designated by the letter "f." The value 300,000,000 is the speed of light of a radio wave in meters per second in free space. (This translates to a speed of 186,410 miles per second.)

A radio signal with a frequency of 150 MHz is expressed in hertz as 150,000,000 Hz. Therefore,

$\lambda = 300{,}000{,}000/150{,}000{,}000 = 2$ meters

If we give the frequency in megahertz, the formula can be shortened to:

$\lambda = 300/f$

For example, we can calculate the wavelength of the shortwave or high-frequency part of the spectrum (3 to 30 MHz) as follows:

$\lambda = 300/3 = 100$ meters
$\lambda = 300/30 = 10$ meters

If you rearrange the formula, you can also find the frequency if you know the wavelength:

$f = 300/\lambda$

What is the frequency of the 25-meter band?

$f = 300/25 = 12$ MHz

RADIO WAVE PROPAGATION

Propagation refers to the way in which a radio signal travels from the transmitting antenna to the receiving antenna. The way the signal behaves is directly related to its frequency.

Low and Medium Frequencies

For example, in the low-frequency and medium-frequency ranges, the radio signals tend to hug the Earth as they move from transmitter to receiver. These ground wave signals can only travel relatively short distances because they are absorbed and attenuated by the Earth, trees, buildings, and other obstructions. Depending on the frequency of operation, transmitter power, and the nature of the terrain, distances are limited to several hundred miles. Distances can be extended with greater power or if the signal travels over water.

High Frequencies

In the HF or shortwave frequency range, the ground waves are extremely weak and travel only short distances, usually several miles. Instead, propagation is by way of refraction of the ionosphere. The ionosphere is an electrified area above the Earth located from approximately 30 miles to 250 miles above the Earth. The Sun ionizes the gases in the air, thereby giving them electrical characteristics. The closer the ionosphere to the Sun, the greater the degree the ionization. The ionosphere is actually several different layers or ionization levels as shown in Figure 7.5. The upper layers have greater ionization than the lower layers. Obviously, the degree of ionization is determined by whether it is day or night. During the daylight hours, the ionosphere is highly charged electrically. At night, the ion layers essentially disappear.

A radio wave in the 3- to 30-MHz range is transmitted up to the ionosphere where it is refracted or bent by ionization. The degree of bending depends on the frequency of operation and how high in the ionosphere the radio wave penetrates. In most cases, the radio wave is bent so much that it actually returns to Earth. The effect appears to be similar to that of a light beam being reflected from a mirror.

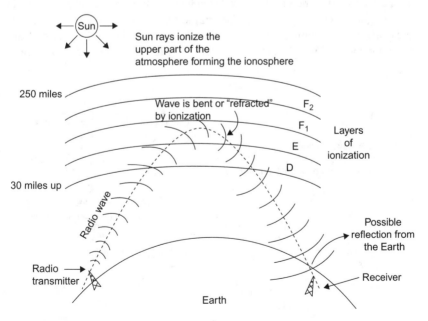

FIGURE 7.5 Radio propagation of shortwaves.

As you can see, the radio wave actually returns to the Earth at some distance from the transmitter. Receiving stations in the area of the refracted signal can receive it. In some cases, if the signal is strong enough and the characteristics of the Earth are right, the radio signal may actually be reflected by the Earth back up to the ionosphere where it is again bent and returned to Earth. Under the right conditions, the radio signals can make several hops, thereby greatly extending the distance of communications. With the optimum frequency, power level, and time of day, communications halfway around the world can easily be achieved. That is why it is easy for you to listen to worldwide radio broadcasts that occur in the shortwave frequency bands.

VHF, UHF, and Microwaves

Radio wave propagation in the frequency range above 30 MHz occurs essentially by direct antenna-to-antenna space waves. This is usually called *line-of-sight (LOS) communications*. Further, the higher the frequency of the radio wave, the less it is affected by the ionosphere. UHF and microwave signals penetrate the ionosphere, making satellite and long-distance spacecraft communication possible.

LOS communications means that the transmitting antenna must be able to "see" the receiving antenna. If the distance of the receiving antenna is beyond the curvature of the Earth, the transmitted signal will simply pass over the antenna. To receive the signal, the receiving antenna or transmitting antenna can be increased in height, thereby extending the total range of communications. Figure 7.6 shows this idea. The simple formula along with it allows you to calculate the maximum LOS communications if you know the height of the receiving and transmitting antennas. That's why at VHF, UHF, and microwave

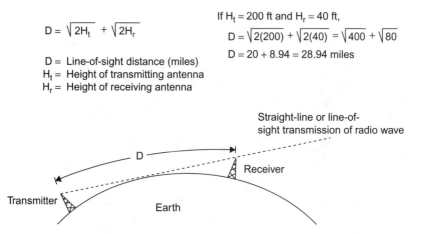

$$D = \sqrt{2H_t} + \sqrt{2H_r}$$

D = Line-of-sight distance (miles)
H_t = Height of transmitting antenna
H_r = Height of receiving antenna

If H_t = 200 ft and H_r = 40 ft,

$$D = \sqrt{2(200)} + \sqrt{2(40)} = \sqrt{400} + \sqrt{80}$$

$$D = 20 + 8.94 = 28.94 \text{ miles}$$

Straight-line or line-of-sight transmission of radio wave

D

Receiver

Transmitter

Earth

FIGURE 7.6   Line-of-sight (LOS) communications at VHF, UHF, and microwave.

frequencies, the height of the antenna is the most critical determining factor in overall communications distance.

MODULATION

A radio wave is a sine wave of voltage that has been converted into electric and magnetic fields by the antenna. The transmitter generates the sine wave and amplifies it to a high power level. The question is, how does the information to be transmitted fit into this?

The information or intelligence to be conveyed is usually a voice signal produced by a microphone. Other forms of intelligence signals include pictures or video and digital (binary) data from a computer. The purpose of the high-frequency sine wave radio signal is to give the intelligence signal a ride. That's why the radio signal generated by the transmitter is called the carrier. Modulation is the process by which the information is impressed on or embedded in the carrier. Modulation causes the carrier to be modified by the information signal.

There are three basic forms of modulation used to put information on a carrier: *amplitude modulation* (AM), *frequency modulation* (FM), and *phase modulation* (PM).

Amplitude Modulation

In amplitude modulation, the voltage or power level of the information signal changes the amplitude of the carrier in proportion (see Figure 7.7). With no modulation, the AM carrier is transmitted by itself. When the modulating information signal (a sine wave) is applied, the carrier amplitude rises and falls in accordance. The carrier frequency remains constant during amplitude modulation.

Amplitude modulation is widely used in radio. AM broadcast stations are, of course, amplitude modulated. So are citizens band radios, aircraft radios, and the video modulation of a TV broadcast transmitter. A special form of amplitude modulation, known as quadrature modulation (QAM), is also widely used in modems to transmit digital data over cable or wireless.

Sidebands

The modulation process causes new signals to be generated. These new sine wave signals are called sidebands. Their frequencies are the sum and difference of the carrier and modulating signal frequencies. For example, in an AM radio station, audio frequencies as high as 5 kHz can be transmitted. If a 5-kHz sine wave tone is to be transmitted, the modulation process causes sidebands 5 kHz below and 5 kHz above the carrier to be produced. This is illustrated in Figure 7.8. For an AM radio station with a carrier frequency of 860 kHz, the

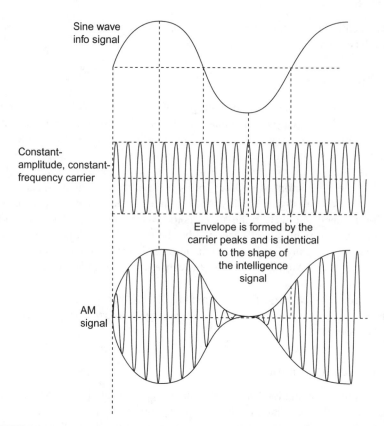

Sine wave info signal

Constant-amplitude, constant-frequency carrier

Envelope is formed by the carrier peaks and is identical to the shape of the intelligence signal

AM signal

FIGURE 7.7   Amplitude modulation.

lower sideband (LSB) would occur at $860 - 5 = 855\,\text{kHz}$, while the upper sideband (USB) occurs at $860 + 5 = 865\,\text{kHz}$. The carrier and the sidebands combined produce the composite waveform shown at the bottom of Figure 7.7.

Bandwidth

This brings up the very important concept of bandwidth. Bandwidth refers to a range of frequencies over which a radio signal operates. This is also referred to as a radio channel. As you can see from Figure 7.8, the AM signal consists of the carrier and the sidebands. Together these signals occupy a bandwidth of 10 kHz. You can figure out the bandwidth of any signal by simply subtracting the lower sideband frequency from the upper sideband frequency.

$$BW = USB - LSB = 865 - 855 = 10\ \text{kHz}$$

All of the receiver circuits must be set to pass signals in this 10-kHz range in order to avoid distortion of the signal or lost information.

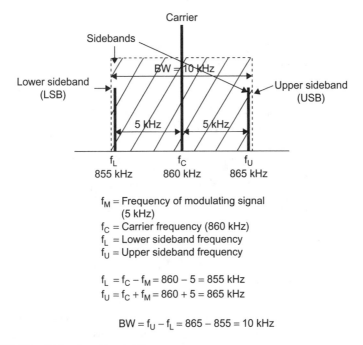

FIGURE 7.8   Sidebands and bandwidth.

Frequency Modulation

In frequency modulation, the carrier amplitude remains constant but its frequency is changed in accordance with the modulating signal. Specifically, the higher the amplitude of the information signal, the greater the frequency change. The actual carrier frequency deviates above and below the center carrier frequency as the information signal amplitude varies. Figure 7.9 shows frequency modulation with a sine wave information signal. Note that the carrier frequency gets higher on the positive peaks and lower on the negative peaks of the information signal.

Like AM, FM also produces sidebands. But unlike AM, which produces a single pair of sidebands for each frequency in the modulating signal, the frequency-modulation process produces an infinite number of pairs of sidebands for each frequency in the information signal. As a result, the bandwidth occupied by an FM signal is enormous. Luckily, the number of sidebands produced can be controlled by properly selecting the amount of deviation permitted in the carrier. Small deviations result in fewer sidebands. Further, some of the higher-order sidebands are extremely low in amplitude and, therefore, contribute little to the FM signal. But while the bandwidth of an FM signal can be controlled and established to fit a desired frequency range, it does nevertheless usually take more room in the spectrum than an AM signal.

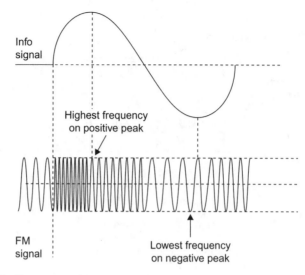

FIGURE 7.9   Frequency modulation.

The primary benefit of FM is that it is less sensitive to noise, which consists of undesirable amplitude variations that get involuntarily added to a signal. Noise is easily eliminated in an FM system where a constant carrier amplitude is used. Some of the most common applications of FM include FM radio broadcasting, and two-way mobile and marine radios.

Phase Modulation

The third type of modulation is phase modulation. Remember that phase shift is a time shift between two sine waves of the same frequency. We can use the information signal to shift the phase of the carrier with respect to the carrier reference. The result of this, however, is a signal that looks virtually the same as an FM signal (Figure 7.9). Phase modulation is usually a little easier to implement electronically than FM, so most so-called FM systems typically use phase modulation instead. Both FM and PM are often referred as types of *angle modulation*.

Digital Modulation

All of the three basic types of modulation, AM, FM, and PM, can also be used to transmit digital or binary data. However, they are usually given different names. Amplitude modulation of a carrier by a binary signal is usually referred to as amplitude shift keying (ASK). This is illustrated in Figure 7.10A. The binary signal simply shifts the carrier amplitude between two specific levels. A special form of ASK is called on–off keying (OOK). This is illustrated in Figure 7.10B. Here the binary signal simply turns the carrier on for a binary 1 and off for a binary 0.

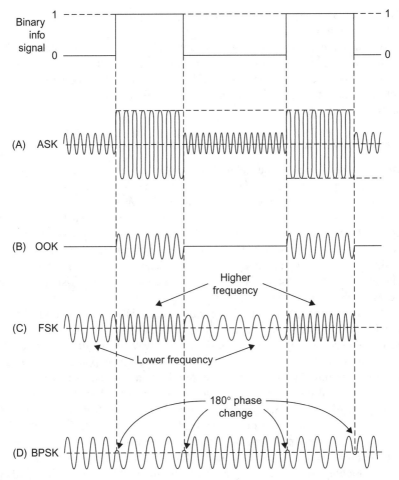

FIGURE 7.10   Types of digital modulation. **(A)** ASK. **(B)** OOK. **(C)** FSK. **(D)** BPSK.

Frequency modulation of a carrier by a binary signal is called frequency shift keying (FSK). Here the binary signal shifts the carrier between two discrete frequencies (see Figure 7.10C).

Phase modulation of a carrier by a binary signal is referred to as phase shift keying (PSK). The term binary PSK (BPSK) is also used. The phase of the carrier is changed as the binary signal switches from 0 to 1 or 1 to 0 (see Figure 7.10D). Phase shift is 180 degrees, which is easily detected at the receiver.

One of the most widely used forms of digital modulation is QAM, which is really a composite of both AM and PSK. The binary signal is converted into a multilevel digital signal that simultaneously modifies both the amplitude and the phase of the signal. QAM is used in digital data transmission in satellites and in modems sending digital data in audio form over telephone lines or cable

TV systems. It is also widely used in the newer digital wireless networks. See Chapter 8 for more detail.

Because binary signals contain many higher-frequency harmonics, the resulting signals produced by modulation have an enormous bandwidth. Many schemes have been developed to limit the spectrum of the modulating signal and, as a result, to reduce the amount of bandwidth occupied by a binary modulated radio signal.

Spread Spectrum

Spread spectrum (SS) is unusual in that it is not only a form of modulation but also a system for encrypting the information signal for security purposes and for multiplexing multiple signals over a range of frequencies.

Spread spectrum was originally developed during World War II as a way to prevent the enemy from interpreting radio transmissions. For years the government and military protected the concepts of SS modulation. However, today it is widely used in radio and telephone communications. Most of the newer cellular telephones are based on this concept, specifically code division multiple access (CDMA) cell phones.

When using SS, the analog information signal is converted into a digital signal before transmission. The digital voice signal is then used to modify the carrier in a special way. The two most common ways of modifying the carrier are referred to as *direct-sequence* (DS) and *frequency-hopping* (FH) *spread spectrum*. In direct-sequence SS, the serial digital signal representing the voice is chopped up into random sequences and then used to PSK the carrier. In frequency-hopping SS, the carrier is rapidly switched from one frequency to another for a short dwell time. The result of SS modulation is that the signal itself is broken up at random, and spectral energy is spread over a very wide frequency range. Because of the random sequences of the coding and frequency hopping, such a radio signal is nearly impossible to put back together at a receiver without knowing the code. SS receivers designed to pick up the SS signal know the code, and can unscramble the signal and reassemble it into the original modulating information signal.

RADIO HARDWARE

All radio communications systems consist of three basic pieces of hardware: *transmitter*, *receiver*, and *antenna*. In some systems, the transmitter and receiver are packaged and used separately, while in others they are combined into a single package and called a *transceiver*.

Transmitters

The radio signal is produced by the transmitter. A general block diagram that applies to virtually every transmitter is shown in Figure 7.11. The radio signal

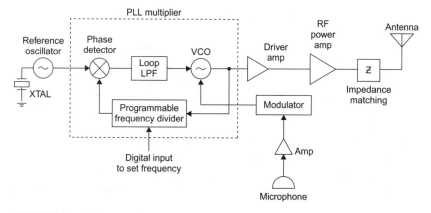

FIGURE 7.11   Radio transmitter.

is generated at a low power level by a frequency synthesizer. This uses a phase-locked loop (PLL) with a crystal (XTAL) oscillator reference. A variable-frequency divider is used to set the frequency of operation. The FCC is really fussy about the frequency at which transmitters work and, therefore, the frequency must be precisely set and extremely stable. A quartz crystal reference oscillator is normally used to achieve this objective.

The low-level signal produced by the voltage-controlled oscillator (VCO) is then amplified by one or more driver amplifiers to boost the power level. Next, the microphone signal is fed to the modulator. The modulator changes the frequency or phase of the RF signal in accordance with the voice, digital data, or other information to be transmitted. That signal is applied to the PLL VCO.

The modulated signal is sent to additional RF power amplifiers, usually referred to as *drivers*. Each succeeding stage boosts the signal power by a specific amount. A final RF power amplifier produces the desired amount of output power. From there, impedance (Z) matching circuits are used to match the high-power amplifier to the antenna for maximum power radiation.

The physical realization of the transmitter can vary all the way from a single integrated circuit chip the size of a postage stamp to something that can fill up a moderately large building. The size depends on the operating frequency and the amount of output power desired. Very-low-power, high-frequency radio transmitters for garage door openers and remote keyless entry systems on cars are fully self-contained within a single integrated circuit. High-power radio broadcast transmitters at low frequencies may fill a small room. The higher the power and the lower the frequency, the larger the transmitter.

Receivers

The job of the receiver is to translate the weak radio signal picked up by the antenna into an output that fully recovers the originally transmitted information.

By the time the radio signal reaches the receiving antenna, it is extremely weak and, therefore, the primary job of the receiver is to boost the signal level. Therefore, the receiver consists mainly of a series of amplifiers that increase the level of the small signal to a point where the original information can be recovered.

The amount of amplification provided by a receiver determines its *sensitivity*. Another job of the receiver is to provide selectivity so that the desired signal can be picked out of the thousands of signals being transmitted simultaneously by other transmitters. The receiver's job is to select just the desired one and to reject all of the others. This is what we refer to as the *selectivity* of the receiver.

A third function of the receiver is to recover the original information. A circuit called a *demodulator* removes the voice, video, or digital data from the carrier. Finally, output amplifiers reproduce the signal as desired, such as a voice signal in a speaker.

Figure 7.12 is a block diagram of a superheterodyne receiver. The incoming signal is usually amplified by an *RF input amplifier* often called a *low-noise amplifier* (LNA) and then fed to a circuit called a *mixer*. The mixer is a frequency-translation circuit that converts the incoming signal, regardless of its frequency, along with its modulation to a lower intermediate frequency (IF). This allows the receiver to use a fixed-frequency amplifier system where most of the gain and selectivity of the receiver is achieved. The mixer takes the input signal and mixes it with a high-frequency sine wave signal generated internally in the receiver by what is called a local oscillator (LO). The LO is usually a frequency synthesizer. The result of the mixing is output signals that are the sum and difference frequencies of the input and oscillator signals. The IF amplifier is normally tuned to select only the difference frequency. Tuning is accomplished by changing the local oscillator frequency. The local oscillator in most channelized receivers is some type of PLL frequency synthesizer.

The IF amplifier boosts the signal level considerably. Its selectivity ensures that signals whose frequencies are above or below the desired signal are effectively eliminated. The IF amplifier also has a built-in feedback control circuit that automatically adjusts its gain to the level of the incoming signal. This feature is called *automatic gain control* (AGC). AGC permits the receiver to pick up both strong local signals and very weak distant signals. When very strong signals are received, the AGC automatically reduces the gain of the IF amplifier to minimize circuit overload and signal distortion. When very weak signals are received, the AGC adjusts the gain of the IF amplifier for higher amplification.

The *demodulator circuit*, sometimes called the *detector*, follows the IF amplifier. It removes the carrier and recovers the desired information signal. From there, additional processing takes place to either boost the signal level or to shape it before it appears at the output.

Most receivers that we are familiar with have a speaker as the primary output. But there are exceptions, such as the picture tube in a TV set or the motor in a garage door opener. The receiver may also put out digital data that goes straight to a computer.

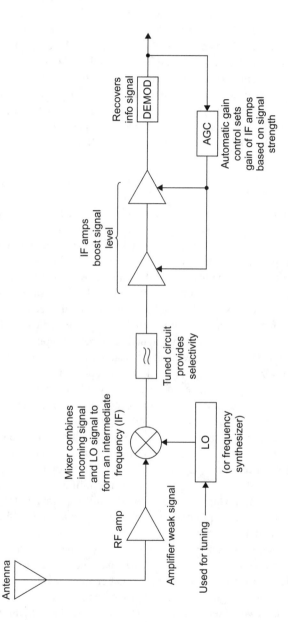

FIGURE 7.12   Superheterodyne receiver. LO, local oscillator.

Transceivers

Most communications systems use both a transmitter and a receiver. In two-way radio, both send and receive capability is needed at both ends of the communications channel. Therefore, the receiver and transmitter are packaged together in the same housing to create a transceiver. Small portable transceivers are called hand-helds, walkie-talkies, or cellular telephones.

The transmitters and receivers are typically the same as those described earlier. However, in some cases, the transmitter and receiver may share circuits. They virtually always share the antenna and a common power supply like a battery. In some cases, a single-frequency synthesizer may generate both the receiver LO signal as well as the transmitter carrier.

ISM Band Transceiver

The ISM band refers to specific frequencies set aside for wireless industrial, scientific, and medical (ISM) applications. The most common frequencies in the United States are 315, 433, and 915 MHz. The frequency 868 MHz is widely used in Europe. These unlicensed frequencies are used for simple monitor and control applications using low-speed digital data over short ranges. Some common examples are garage door openers, remote keyless entry on cars, remote thermometers, tire pressure gauges, security and fire alarms, home automation, and any telemetry or remote control application. The radios used for this are typically single ICs to which an antenna, a battery, a housing, and the input data are added. The devices are available as a receiver, a transmitter, or a complete transceiver. Figure 7.13 is a block diagram of a typical transceiver.

The transmitter section uses an external crystal to serve as the reference to an internal PLL frequency synthesizer to set the frequency. The PLL actually multiplies the crystal frequency by some N value to get the desired output. If the crystal frequency is 6.77 MHz and the divide ratio is 64, the output would be 433.28 MHz. The serial digital data input coming from a sensor, an analog-to-digital converter (ADC), or a computer frequency modulates the crystal oscillator to create the FSK signal. A power amplifier (PA) after the transmit VCO boosts the power up into the 1- to 100-milliwatt range. The output goes to the antenna. A transmit/receive switch allows the antenna to be shared in a half-duplex system.

The receiver is a special version of a superheterodyne. The input signal from the antenna is boosted in amplitude by the LNA and applied to the mixers. The local oscillator is set to the incoming signal frequency (also the transmit frequency), so the mixers generate a difference of zero. This is called *direct conversion* or *zero IF*. Two mixers are used and the local oscillator is shifted 90 degrees in phase between the two. This gives two zero IF outputs that are filtered. Zero IF does not mean a nonexistent output signal. It just means that the FSK signal is recovered directly at its data rate. The I and Q signals are 90 degrees out

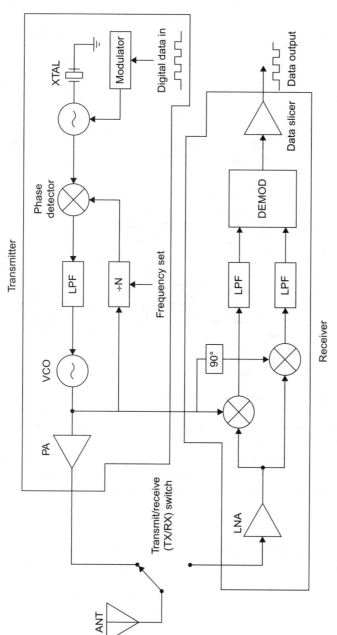

FIGURE 7.13 Single-chip UHF transceiver for low-speed, short-range digital data transmissions.

of phase and this allows the digital data to be more easily demodulated. A data slicer, a circuit like a comparator, is used to shape the signal into a clean binary bit stream. With this kind of radio, the range is limited. It may be only a few feet or up to a mile or so depending on the antenna, environment, or application.

ANTENNAS

The antenna takes the voltage and current produced by the transmitter and converts it into the electromagnetic wave we know as a radio wave. At the receiver, the antenna converts the electromagnetic field back into a voltage we know as the radio signal. The radio waves pass over the receiving antenna, inducing a voltage that causes current to flow in the receiver.

Antennas can have an amazing variety of shapes. Essentially an antenna or aerial is some arrangement of electrical conductors whose size and length are critical. For optimum operation, the antenna length must be related to the wavelength (λ) of the signal. Common antenna sizes are one-quarter ($\lambda/4$) and one-half ($\lambda/2$) wavelength at the operating frequency. The higher the frequency, the shorter the wavelength and thus the smaller the antenna. For example, one wavelength at 150 MHz, a popular two-way radio frequency range measures 2 meters. A half wavelength is 1 meter and a quarter wavelength is 0.5 meter. Since 1 meter is 39.37 inches, a half wavelength is 19.68 inches. At 900 MHz, the frequency range of cell phones, one wavelength is 0.3333 meter. A quarter wavelength antenna would be $0.3333/4 = 0.08333$ meter or 3.28 inches.

Ground-Plane Antenna

Antennas can be anywhere from a long horizontal copper wire to a vertical steel or aluminum tower several hundred feet tall. Or it may be an array of short conductors like those in a TV antenna. One of the most common antennas is a short quarter wavelength vertical that is widely used in mobile radios, cell phones, and other hand-held units (see Figure 7.14). The quarter-wave or ground-plane antenna as it is sometimes called, radiates equally well in all directions, making it omnidirectional. We say that its radiation or reception pattern is a circle.

Dipole

Another common antenna type is the dipole, which is one-half wavelength long and usually mounted horizontally. Figure 7.15 shows an example. Its radiation or reception pattern is a figure 8. The greatest radiation or strongest reception occurs at a right angle to the dipole element itself.

Yagi

A variation of the dipole is the Yagi, a multielement antenna based on the dipole shown in Figure 7.16. To the dipole is added a reflector element that is

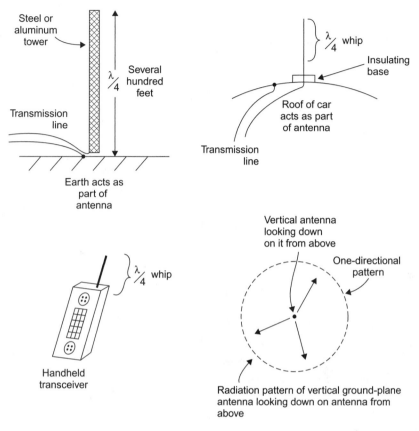

FIGURE 7.14   Quarter-wave ground-plane antenna.

slightly longer than one-half wavelength, and one or more director elements that are progressively shorter than the dipole. The directors and reflector make the Yagi highly directional. The extra elements, called *parasitic elements*, concentrate and aim the beam of radiation over a narrow range. At the transmitter, this has the effect of increasing the transmitted power, making the signal stronger at the receiver as if the transmitter power had been increased. A similar gain effect occurs at the receiver.

Other Antennas

There are many other antenna variations. Antennas formed with copper patterns on a printed circuit board (PCB) are widely used. One example is the *patch antenna* (see Figure 7.17). It is simply a square area of copper about one-half wavelength on a side made on a PCB. Another is the *loop antenna* that is just a loop of copper on the PCB. A popular cell phone antenna is the

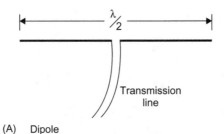

(A)   Dipole

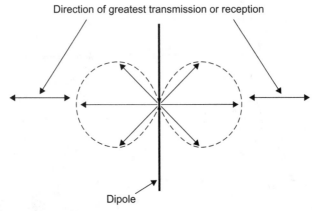

(B)   Horizontal radiation pattern

FIGURE 7.15   Dipole antenna.

inverted F, which is an F-shaped pattern on the PCB. Antennas also come as components. For microwave frequencies, there are special antennas made on a ceramic base that are small and can be soldered to a PCB.

Transmission Lines

The antenna is usually connected to the transmitter or receiver by way of a coaxial cable (see Figure 7.18). It consists of a copper center conductor surrounded by an insulator, such as Teflon. Around that is a shielded copper braid or foil. The whole thing is covered with a plastic insulator. Coaxial cable is the most commonly used transmission line to connect the equipment to the antenna. It provides an efficient method of transferring the voltages and currents to and from the antenna while providing a certain amount of shielding from noise. For PCB antennas, the transmission line is just a short pair of copper lines from the transceiver IC to the antenna.

Microwave Antennas and Waveguides

Antennas used at microwave frequencies are quite different from those used at the lower frequencies. Perhaps the most common is the *horn antenna*. It looks

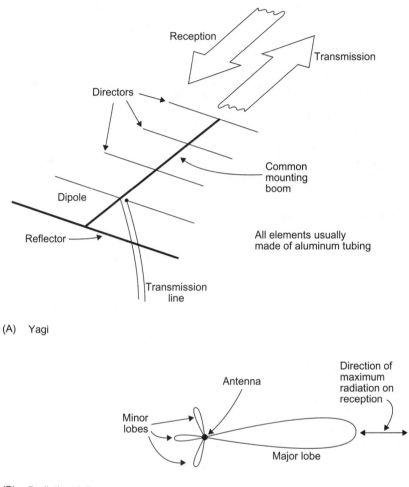

(A)   Yagi

(B)   Radiation pattern

FIGURE 7.16   Yagi antenna.

like the flared end of a musical horn, only most of them have an opening or aperture that is rectangular rather than round.

As seen in Figure 7.19, the horn is directional and has gain. The microwave energy is fed to the horn by a transmission line. Coax can be used for the lower microwave frequencies, but at the higher microwave frequencies, coaxial cable loss is too high for long runs. A high percentage of the signal power is lost in the coax cable itself. This problem is overcome by using a waveguide. A waveguide is nothing more than a hollow copper pipe, usually with a rectangular cross-section through which the microwave energy passes from transmitter to antenna or from antenna to receiver.

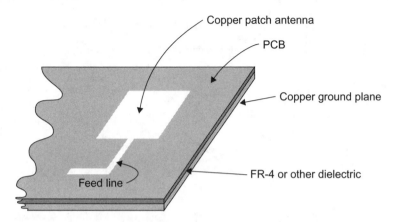

FIGURE 7.17 Printed circuit board (PCB) patch antenna.

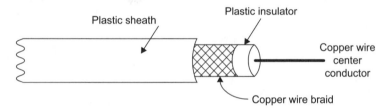

FIGURE 7.18 Coaxial cable transmission line.

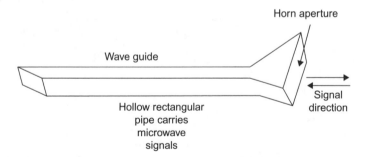

FIGURE 7.19 Microwave horn antenna fed by waveguide transmission line.

The most effective and widely used microwave antenna is the parabolic dish (see Figure 7.20). This is usually a horn antenna combined with a reflector shaped like a dish. Its shape is a mathematical parabola. Signals transmitted by the horn are reflected forward by the parabolic dish shaping the radio waves into a very narrow beam. Received energy is reflected by the dish into the horn antenna. The highly directional nature of the dish gives it a very

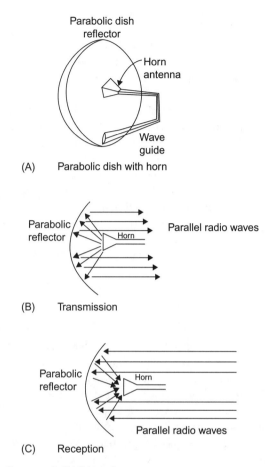

(A)   Parabolic dish with horn

(B)   Transmission

(C)   Reception

FIGURE 7.20   How a parabolic dish works.

narrow beamwidth and enormous gain. These are widely used in satellite TV and radar.

TWO-WAY RADIO

Two-way radio communications between individuals are by far the most widespread use of radio. And there are many forms of it. You have seen such radio communications as they take place and you have no doubt engaged in some form of it yourself. In any case, this section takes a look at some of the diverse uses of such two-way communications.

Most two-way radios operate in the VHF and UHF bands. The simplest two-way radios operate on multiple frequencies usually selected by a frequency synthesizer. Virtually all two-way radio communications are half duplex, meaning that one party transmits while the other receives.

Aircraft

All airplanes from the smallest single-prop planes to the largest jumbo super jets use two-way radio to communicate with airport towers and the major air traffic management centers across the country. Two-way radio is absolutely essential to the safety of any flight.

Most aircraft radios are transmitters installed in front of the pilot or co-pilot. They usually operate in the 108- to 135-MHz band and use amplitude modulation (AM). Specific channels are assigned for different regions and for different communications functions. The pilot or co-pilot uses front panel switches to select the desired communications frequency. Hand-held microphones with a push-to-talk switch are the most commonly used input devices, but other forms of microphones such as throat mikes or microphones attached to headsets are also widely used. Most transceivers operate at very low power and transmission distances are considerable. The height of the airplane gives the radio transceiver an enormous transmission range even with low power at VHF frequencies.

Citizens Band

You have no doubt heard of citizens band (CB) radio and perhaps even used one yourself. Anyone can buy and use a citizens band radio for two-way communications. The FCC has allocated 40 frequencies or channels in the 27-MHz frequency range for citizens band operation. The maximum transmission power is restricted to 5 watts, although some hand-held units use less. AM is used.

Citizens band radio has enjoyed wide popularity with individuals and sportsmen as well as truckers. While the advent of the cellular telephone has reduced the number of individuals using citizens band radio, it is still widely used. The primary limitation of most CB radios is the antenna. Since CB radios operate in the 27-MHz range, the required minimum antenna length is approximately 9 feet. This can be shrunk somewhat by the use of coils attached to the antenna, but effectiveness is reduced. A CB radio antenna of 5 or 6 feet is common and certainly manageable on a car, truck, or boat. In hand-held transceivers, any antenna longer than about 18 inches is a nuisance.

A 27-MHz frequency allocation of CB radio also gives it some peculiar transmission characteristics. While communications is reliable over short distances of up to several miles, at certain times of the day or evening, worldwide communication is also possible. This is probably more a disadvantage than an advantage, primarily because interference can be received from stations literally anywhere in the world.

Amateur Radio

Amateur or ham radio is a hobby. Technically knowledgeable individuals are licensed by the FCC to use selected frequency bands across the spectrum for personal communications. Hams use voice as well as CW (continuous wave) or Morse code transmissions. Hams talk to one another about their equipment, the weather, and other common matters. Communication is worldwide and helps to promote good international relationships. Hams frequently design, build, and experiment with their own equipment, although many hams prefer commercially manufactured radios. Hams experiment with digital radios, satellites, and TV.

Family Radio

Family radio is another radio service recently instituted by the FCC for the purpose of short-range personal communications. It supplements CB radio in a wide range of personal communications needs. Because of a different frequency allocation, family radios are smaller, more portable, and have more reliable communications over distances of several miles.

All family radios are hand-helds using FM and operating in the 450-MHz range. Only several frequencies in the 450 MHz band are allocated to the Family Radio Service. While many users will share these frequencies, the short communication distances minimize interference between stations using the same frequency. The transceivers are restricted to 2 watts in power. At a frequency of 450 MHz, a quarter-wave whip antenna is only about 6 inches long. Communications are extremely reliable up to 1 mile and even farther if transmission takes place outdoors.

You can use family radios in a variety of ways. They are particularly handy when you are camping, hiking, or doing other outdoor activities. Many families use them in shopping malls and even around the house to keep in touch. They are great on farms and ranches and for communications between two nearby cars.

Marine

The very first application of two-way radio in the world was on ships at sea. Ships were always in danger without any form of communication. If a ship was disabled or began sinking, no one would know for a long period of time or maybe never. With a two-way radio or even a one-way radio on a ship, the operator could communicate with a shore-based radio station to signal danger with the familiar SOS.

Today, a radio is a necessity on virtually every boat and ship. While small boats are not required to carry radios, most larger boats are. And what seaman would be without a radio for safety reasons?

Marine radios have operated on a wide range of frequencies over the years. Ocean-going liners have operated on shortwave frequencies so that long-distance communications are possible. Ships also operate on lower frequencies for exchanging messages and at the higher frequency for reliable communications near land and in port.

The typical marine radio for communicating near ports is a transceiver of 25 to 40 watts of power operating with frequency modulation in the 150- to 170-MHz range. Most of these units transmit on multiple frequencies, making it always possible to find a clear channel. Channel 16, operating on a frequency of 156.8 MHz, is used for emergency calls. This frequency is typically monitored by other boaters and always by the Coast Guard.

Mobile Radio

Mobile radio refers to two-way radios mounted in cars and trucks. These are widely used by police, fire, and other civil services. They are also widely used by taxis, buses, and by companies using fleets of cars or trucks for convenient communications. The typical unit is mounted in the vehicle and provides relatively high power in the 25- to 50-watt range. Most communications take place in the 150- to 170-MHz range. Other mobile radio frequency ranges are 450 to 470 MHz. Most radios today are hand-held types.

SATELLITES

A satellite is a space station that is designed to orbit the Earth and provide a remote radio signal relaying station for the purpose of extending the range of communications worldwide. See Figure 7.21. Such communications satellites contain both a receiver and transmitter that operate simultaneously. The receiver picks up signals from the Earth on one frequency and retransmits them simultaneously on another frequency. In this way, the transmitter and receiver in the satellite do not interfere with one another.

Most communications satellites are geosynchronous. These are satellites positioned to orbit directly around the Earth's equator at a distance of 22,300 miles from Earth. When located in this position, the satellite rotates in exact synchronization with the Earth's rotation. Thus, the term *geosynchronous*. Because the satellite and the Earth are rotating together, it appears as though

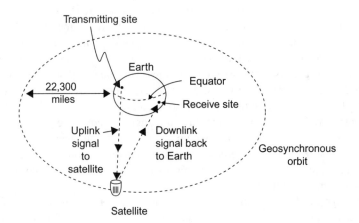

FIGURE 7.21   Concept of communications satellite.

the satellite is stationary overhead to an observer on Earth. With the satellite in a fixed position with respect to the Earth, the satellite becomes a near-perfect radio repeater station.

Figure 7.22 shows a general block diagram of a satellite. The communications part of a satellite consists of a combined receiver and transmitter called a transponder. Satellites operate in the microwave region. The C band from 4 to 6 GHz and the K_u band from 11 to 18 GHz are the most popular operating ranges. A low-noise amplifier (LNA) picks up the very small 6 GHz uplink signal from Earth. It is mixed with a 2-GHz LO signal and down-converted to a 4-GHz downlink signal, which is then amplified by a high-power amplifier (HPA) and retransmitted back to Earth. Most satellites contain many transponders with wide bandwidth capable of handling many signals simultaneously.

The satellite power source is batteries combined with a system of solar cells. The solar cells convert light from the Sun into electrical energy that is used to recharge the batteries. This system gives the satellite long life in space.

Communications satellites are used primarily in long-distance telephone communications and for distribution of TV signals. Some satellites are used for navigation, a good example being the global positioning system (GPS) widely used by the military and commercial organizations. It is also widely used by individuals as they find their way with personal navigation devices (PNDs) like those from Garmin and Tom Tom. The GPS system uses 26 satellites about 10,800 miles above the Earth orbiting in overlapping patterns so that at least three of four are "visible" by wireless at any point on the Earth.

Other satellites are used for surveillance. Spy satellites only 102 miles high pick up radio signals as well as infrared images and photographs. Weather satellites monitor cloud formations and movements. There are also two satellite telephone networks with low-Earth-orbit satellites to permit communications to/from any point on the planet.

RADIO TELESCOPES

Observation of the planets and stars is usually done with an optical telescope. However, in 1932 researchers discovered that distant stars or suns emitted radio waves. All stars, like our Sun, emit radio frequencies. Most of this energy occurs in the microwave region at 1.4218 GHz. This is the frequency of ionized hydrogen. By building a highly sensitive radio receiver and coupling it to a large highly directional radio antenna, it is possible to plot the position of distant stars just as effectively as an optical telescope. Radio telescopes are, in fact, even more effective in that they can more easily detect distant stars that cannot be seen even by the best optical telescopes.

Radio astronomy observatories consist of an enormous parabolic dish antenna usually 60 to 250 feet in diameter, which can be rotated and tilted so that it scans the skies. An array of multiple smaller dishes can also be used as in the SETI (search for extraterrestrial intelligence) project in New Mexico. The antennas are coupled to one or more very sensitive radio receivers. These

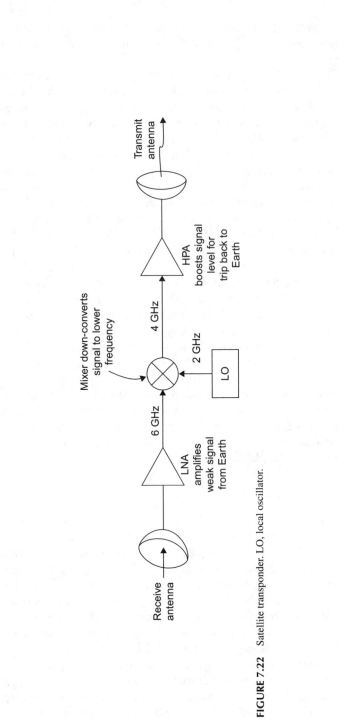

FIGURE 7.22   Satellite transponder. LO, local oscillator.

receivers have very high gain so that they can sufficiently amplify very small signals received from stars that are many light years away. Often special cooling systems are used to reduce the temperature of the receiver front end to minimize noise, thereby permitting even weaker signals to be received. The largest radio telescope is a parabolic dish reflector built of screen mesh 1000 feet in diameter. The dish is built into the valley between adjacent mountains near Arecibo, Puerto Rico. The horn antenna is moveable on a long cable so it can be positioned to see any part of the sky.

Wireless Everything

It seems like everything is going wireless these days. Cordless telephones, wireless local area networks, and your TV remote control. But wireless has always been around just not by that name. Most radio applications are old and well known. A few examples you may have encountered follow.

Garage door opener. A tiny transmitter in your car transmits a binary coded message to the receiver in your garage hooked up to the door motor.

Remote keyless entry. A miniature transmitter on your key ring transmits a signal to a receiver in your car that opens the door locks. Some even turn on lights and start your car from 10 to 30 feet away.

Wireless data acquisition. Remote transmitters are set up with solar batteries at remote sites to gather data from sensors and transmit it back to a collection center. A good example is electric, gas, and water utilities that use such systems to monitor critical conditions at remote sites and transmit the data back to a central monitoring facility where decisions can be made. Home wireless thermometers are very common.

Radio control of models. This is very common in toy cars, model airplanes, and boats.

RFID. Ultra miniature radios are being embedded in shipping labels, cars, and other objects for identification purposes. The units transmit their unique ID code when scanned by a nearby receiver.

HD radio. This is digital radio broadcasting in the United States. Most U.S. AM and FM stations now transmit a digital version of their signals on the same frequencies as their analog transmissions. They use a form of digital modulation called OFDM (see Chapter 8). It produces greater frequency response than the regular analog broadcasts, making AM sound more like FM and FM sound more like a CD. A special radio is needed to receive it. The digital technology makes noise and fading less of a problem. And it permits each station to add several additional channels thanks to the multiplexing capabilities of the HD technology.

Security tags. The plastic tags clipped on new clothes to prevent their being stolen contain a tiny antenna and electronic circuit that make their presence known at a nearby transmitter/receiver unit.

Surveillance and bugging, or secret monitoring or electronic spying. This is not just something you see in the movies or on TV. It is more widely used in private investigation, police work, and industrial espionage than you may think. Scary. Face it, wireless is truly everywhere.

Project 7.1

Investigate HD Radio

HD radio is digital radio in the United States that shares the same frequencies as the regular AM and FM bands. You can buy stand-alone HD receivers, and some stereo receivers include it. HD radio is also available in some cars. To learn more about this technology, go to the website of the developer iBiquity or www.ibiquity.com. Also try a Google/Bing/Yahoo search on HD radio.

Project 7.2

Become Familiar with ISM Radio

ISM radio is the term used to refer to short-range data radios used in the FCC's unlicensed spectrum designated for industrial-scientific-medical (ISM) applications. Got to the FCC website www.fcc.gov and find the most commonly used ISM frequencies of operation in Part 15 of the rules and regulations. Do a general search of ISM to determine some common applications.

Project 7.3

Radar

Radar (radio detection and ranging) is a type of wireless. It has many uses, mainly military but also commercial and civilian. Marine and aircraft navigation are key applications. Police radar is widely used. Radar is also showing up in automobiles as detectors of distance between vehicles for automatic braking. You may wish to learn more about this technology. Do a search on radar.

Project 7.4

Shortwave Radio

Buy a low-cost shortwave radio and experience listening to international broadcasts and amateur radio operation.

Project 7.5

Experience CB or FRS

Go to your local Radio Shack and buy a pair of Family Radio Service walkie-talkies. Then test them to determine maximum range. As an alternative, buy a CB radio and listen to what is going on with the truckers or other locals.

Chapter 8

Cell Phones

It Is Now Possible to Do Anything Wirelessly: Talk, Text, Email, Web Browse, Games, Whatever

> **In this Chapter:**
> - How the cellular system works.
> - Cell phone radio technologies.
> - Digital modulation and access schemes.
> - Cell phone data capabilities.
> - Smartphones, femto cells, MIMO, and location technology.

INTRODUCTION

The cell phone is one of the coolest, most useful, and convenient electronic devices of all time. It also happens to be one of the most complex technically, not only internally but because it is part of the largest communications network in the universe, the telephone system. The reason for a separate chapter on this topic is not only because of the advanced wireless technology involved, but also its importance to all of us personally and to the national economy. Since practically everyone has a cell phone today, it plays a huge role in our lives. And for you techies out there, it is worthwhile knowing a bit of the details of how phone calls work, how text messages get sent, and how you can now access the Internet via your cell phone.

CAN YOU HEAR ME NOW?

We've all said that before. Why? For one simple reason: cell phones are two-way radios with all the usual faults. The cell phone industry usually refers to itself as the wireless industry, but it's all the same thing—radio. Your cell phone contains a *transmitter* and a *receiver* that collectively are called a *transceiver*. You will also hear them called *handsets*. People in the industry also call them *mobile terminals*. Whatever. Because cell phones have to be so small to be practical, they are limited in the amount of power they can transmit and the type of antenna that is practical. Battery drain is a major factor in cell phone design, as you want to get as much talk time as you can on a battery charge. But by limiting the transmit power you automatically minimize the distance

doi: 10.1016/B978-1-85617-700-9.00008-4

over which the unit will transmit. In any case, communications range is related to power as well as antenna type and orientation. And the typical range of a cell phone is a mile or so, depending on conditions. Take a look at the sidebar to review some of the facts about wireless and radio covered in Chapter 7.

Factors Affecting Range and Reliability to Radio Transmission

Range—The greater the distance between any transmitter and receiver, the smaller the signal. Signal strength varies inversely by the square of the distance between transmitter and receiver.

Power—The higher the power the greater the transmission range. More power translates into increased communications distance up to a point.

Antenna gain—Some antennas focus the signal in a specific direction. Doing so gives the antenna *gain*, meaning it multiplies the power of the transmitter and the signal strength at the receiver.

Environment—Cell phones and any other microwave wireless device always transmit farther in clear open space. The ideal is direct line of sight (LOS), meaning that the two antennas can "see" one another. While cell phones don't have to have such a clear path to work, they always work better if they do. Environment also means transmitting from inside a building. Yes, radio waves do pass through walls and other obstructions but as they do they lose considerable strength as they encounter obstacles. Walls, ceilings, floors, and any other obstructions greatly attenuate (make it smaller) the radio signal. So if you are inside a car, shopping mall, office, or airplane, your signal strength loses over half its strength and that severely restricts your range.

Height—Getting your cell phone to the highest possible point will greatly increase transmission distance. This helps the line of sight effect so important to reliable transmission. Getting higher typically isn't practical, but almost nothing helps more than height, especially outside.

Receiver sensitivity—This means how much gain the receiver has. High gain translates into greater range.

Multipath—Whenever you are transmitting in an environment with lots of obstructions, your signal will bounce off many objects and be reflected. Some reflected signals will get lost and never get to the receiver. Others will bounce off objects that will ultimately direct them to their intended destination. These reflected signals will be delayed in time. Depending on factors such as operating frequency and distance traveled, these signals could arrive in phase and actually boost reception level. Or they could arrive out of phase and literally cancel the more direct signal that arrived earlier.

Noise—Don't forget that noise also affects communications distance and reliability. See discussion in Chapter 7.

THE CELLULAR CONCEPT

The original designers of the cell phone created a system that helps overcome the range and power problem described in the sidebar. If distances are kept

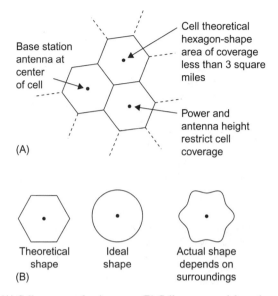

Base station antenna at center of cell

Cell theoretical hexagon-shape area of coverage less than 3 square miles

Power and antenna height restrict cell coverage

(A)

Theoretical shape

Ideal shape

Actual shape depends on surroundings

(B)

FIGURE 8.1 **(A)** Cell coverage of a given area. **(B)** Cell area covered depends on antenna pattern and environment.

short, then other limitations in power and environment are offset. The working word here is "cell." A cell or cell site is one of many transmit/receive stations set up to communicate with individual cell phones. Also known as base stations, these facilities are easily recognized by their tall antenna towers. The small building at the base of the tower houses racks of transceivers that share the big antennas at the top of the tower. Long coax cables carry the signals to and from the antenna. They actually look like pipes going up the side of the tower.

Cell phone systems have dozens to hundreds of cell sites. An ideal design is shown in Figure 8.1A. Each cell represents an area covered by the base station antennas. Cell range and overall coverage are deliberately restricted by antenna height and radiation pattern as well as transmitting power. The hexagon-shaped cells are only theoretical. In reality, the shape of the cell is more circular or rather an irregular circular shape because of antenna characteristics and environmental characteristics that affect the cell shape (nearby trees, buildings, etc.) (see Figure 8.1B).

While equal-size cells are desirable and operation more predictable, in the real world, the coverage of each cell depends a great deal on the number of cell phone subscribers as well as the terrain and environment. In large cities with lots of people and buildings, smaller cell sizes are used to increase subscriber capacity and provide the most reliable coverage possible in a given area. Many cities have what are called microcells that may only cover one block or picocells that cover one area of a building. On the other hand, in rural areas and along

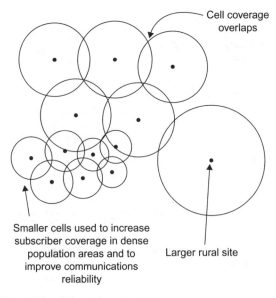

FIGURE 8.2   How small and large cells cover an area.

major highways, the cell sizes are larger and cover fewer users. The cell coverage may look more like that in Figure 8.2.

A key feature of the cellular concept is that by keeping the cell sizes small, the frequency channels assigned to cellular service can be reused. One frequency may carry different calls if the cell sites are far enough apart and do not interfere with one another. This frequency reuse concept multiplies the overall number of channels available to handle calls.

One other factor to keep in mind is that in most cities, there are two or more cellular companies vying for your business. Each company has its own cellular system with base stations with coverage that overlays the cell coverage of other companies. The competing systems do not interfere with one another because they usually operate on different frequencies with different radio technology, and the cell phones they support know which base stations to talk to.

One other thing: when you are talking on your cell phone, you are talking to only one base station—typically the one closest to you since both the power and range of your handset and that of the base station are limited. If you are moving (such as in a car), you will eventually leave the coverage area of one cell and enter the coverage area of an adjacent cell. You never really know when this happens, but it is a pretty neat technical trick to make this happen. The process of having your call transferred from one cell site to another automatically is called the *hand-off*.

All of the base stations are tied to a central office known as the *mobile telephone switching office* (MTSO). Many companies simply refer to it as the *switch*, as its main job is switching between base stations as well as linking to

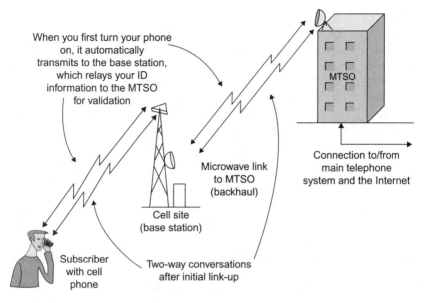

When you first turn your phone on, it automatically transmits to the base station, which relays your ID information to the MTSO for validation

MTSO

Connection to/from main telephone system and the Internet

Microwave link to MTSO (backhaul)

Cell site (base station)

Subscriber with cell phone

Two-way conversations after initial link-up

FIGURE 8.3   Cell phone basic operations.

the main telephone system. The connection between the base stations and the switching station can be by copper cable or fiber optic cable or in many cases by a separate microwave wireless link. This is known as *backhaul*. If a cell site tower has one or more small "dish" antennas on it you can bet that your signal eventually is communicated back to the switch via another wireless link.

The big job of the MTSO is to control and keep track of everything. It validates you as a subscriber when you turn on your phone and make a call. It keeps track of how many minutes of time you use. And it will get messages from remote switches when you are in the roam mode outside of your usual home subscriber area. The cell phone companies have the ability to use other systems in most major cities and in many foreign countries, so you can use the phone virtually anywhere. Figure 8.3 shows just what goes on when you turn on your phone and make or receive a call.

WHAT'S INSIDE A CELL PHONE?

A cell phone can legitimately be called the world's most complicated two-way radio. Sure, it is just a transceiver, but it uses practically every radio trick in the book to ensure that you can make your calls reliably anytime with minimal hassle. It is a two-way radio that is used just like any other phone you are familiar with. In other words, you do not have to say "over," "come back," or any other phrase when you are finished talking and want to listen, as with some two-way radios (CB, family radio, ham radio, aircraft, marine, etc.).

Simultaneous send and receive is referred to as *full duplex*. All telephones are full duplex. It is easy to do in the wired telephone system but complex and costly to do by radio. Yet, full duplex is what makes cell phones so comfortable to use. There is none of this "roger and over" stuff. What this means is that your transmitter and receiver are working at the same time. There is more about how duplexing works later.

Figure 8.4 shows a general block diagram of what's in a cell phone. The receiver and transmitter making up the transceiver are obvious and they share the single antenna. The transceiver is usually a single chip. What is not so obvious is what we call the baseband part of the cell phone. Baseband refers to the voice and data to be transmitted or received. In most modern digital cell phones, this is a very complex integrated circuit that handles the translation of voice between analog and digital, modulation and demodulation, and voice compression. All these functions are done digitally by a digital signal processor (DSP) or special DSP circuits in combination with a powerful embedded processor. This processor or a separate processor handles all the transparent housekeeping jobs of managing the keyboard and the LCD screen as well as the basic control features of the cell system, which also include automatic operating frequency selection and automatic power control under the direction of the MTSO via the cell site. The DSP also implements the messaging functions.

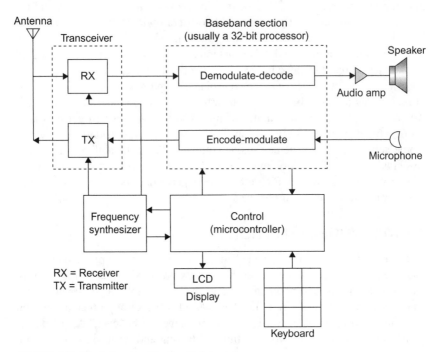

FIGURE 8.4   Block diagram of basic cell phone.

What Type of Cell Phone Do You Have?

Shame on you if you don't know. Well, not really. Actually, it is tough to know just what type of cell phone you have. You can't tell by looking, even inside, since even if you open up your cell phone, God forbid, all you see is a group of integrated circuits, and you can't tell one from another. Furthermore, your cell phone carrier doesn't tell you and usually your user's manual never really says. You can probably ask by calling your carrier, but even then you may not get an answer. I guess the bottom line here is, do you really care? If it does what you want, what difference does it make what technology you are using? On the other hand, if you are an electronics geek, you want to know. I do.

So even though no one knows what type of phone you are using, it does impact the overall system and what phone features and capabilities you have. Multiple types of cell phone technologies are in use concurrently, and that is what makes the cell phone system so bloody complex and expensive. Most of the original carriers have to continue to support the older original technologies as well as implement new ones to stay competitive with other carriers offering their hot new data technologies and features such as email, instant messaging, games, Internet access, mobile TV, and the like.

The first characteristic of a cell phone is the frequencies it uses. Most cell phones operate in the 800- to 900-MHz band where there are hundreds of channels for cell phone calls. Another range of frequencies is the so-called personal communications system (PCS) band from 1850 to 1990 MHz (1.85 to 1.99 GHz), which has hundreds more channels. Some of the newer phones use the advanced wireless services (AWS), 1700- and 2100-MHz spectrum assignments. Because of the higher frequency, PCS and AWS systems have a shorter range, and thus use smaller area cells, which means many more cells sites to cover a given area. Some cell phones actually operate on two or more bands.

Next, a cell phone is known by the technology and access methods it uses. The early original analog phones used first-generation (1G) technology, or FM, but have now been phased out so that only second-generation (2G) digital phones are in use. We are now in the third generation (3G) of cell phone technology and 4G is almost with us. A summary of these technologies follows.

GSM—The Global System for Mobile Communications (originally called Groupe Spécial Mobile) is the 2G digital system developed in Europe. Now all of Europe, most of the United States, and the rest of the world use this excellent but seriously complex technology. It uses time-division multiple access (TDMA), a digital system using digital modulation and a method of multiplying the number of channels in a given amount of frequency spectrum. Your voice is converted to binary 1's and 0's before being transmitted. This allows the carrier to put eight subscribers in one 200-kHz-wide frequency channel, thereby greatly multiplying their subscriber capacity and income. GSM uses a form of FSK called Gaussian minimum shift keying (GMSK). Today, U.S. carriers ATT Wireless and T-Mobile use GSM. By going to

GSM, these carriers can take advantage of several enhancements that facilitate high-speed data transmissions such as the General Packet Radio System (GPRS) and Enhanced Data Rate for GSM Evolution (EDGE).

GPRS—This is an extension of GSM that gives users an always-on packet data transmission capability that they can use for email, Internet access, messaging, or games. GPRS steals one or more of the eight TDMA channels to transmit or receive data instead of digital voice. You can achieve a data rate of about 20 to 60 kbps. GPRS is known as a two-and-a-half generation (2.5G) technology, and is generally available in most larger cities.

EDGE—This 2.5G system is a software upgrade from GPRS/GSM systems that gives you even higher packet data rates. It uses a multilevel modulation called 8PSK that triples the speed of data transmission. Practical speeds up to 180 kbps are possible.

CDMA—Code division multiple access is another 2G digital cell phone technology developed by San Diego–based Qualcomm Inc. The basic technology is known as *spread spectrum* where many signals are transmitted simultaneously over a very wide 1.25-MHz frequency band without interfering with one another. The original standard is called IS-95, and CDMA signals coexist in the same spectrum with TDMA systems in many areas.

cdma2000—This is the 2.5G version of CDMA. It adds high-speed packet data transmission called 1xRTT with a data rate up to 144 kbps. A more recent upgrade called 1xRTT EV-DO has even higher data rates up to 2 Mbps, making Internet access a cinch. Many cdma2000 systems are in operation around the United States by Verizon and Sprint and in Korea.

WCDMA—This is wideband CDMA, which is a third-generation (3G) technology. The access technology is CDMA, but it uses wider 5-MHz bands to provide greater user capacity and very high potential data rates from 384 kbps to 2 Mbps. WCDMA is the upgrade from GSM/EDGE. Carriers using GSM upgraded their systems to WCDMA for greater data speeds and services at greater subscriber capacity. Base stations support both GSM and WCDMA.

HSDPA/HSUPA—High-speed downlink packet access (HSDPA) and high-speed uplink packet access (HSUPA) are upgrades from the WCDMA system. Both use QAM to boost data speeds in the same 5-MHz channels. Data rates as high as 14 Mbps down-link and 5 Mbps up-link are possible. Most cellular companies have now implemented this technology. More advanced versions called HSPA or HSPA+ can give even higher data rates.

cdma2000 Rev A and **Rev B**—These are more advanced CDMA versions that are used to boost data speeds in cdma2000 handsets. Rev A boosts download data rates to 3.1 Mbps using QPSK and 16-QAM in the standard 1.25-MHz channel. Rev B is not widely implemented, but can use three 1.25-MHz channels and boost download speeds to 14.7 Mbps.

Long-Term Evolution—Long-Term Evolution (LTE) is the next or fourth-generation (4G) cell phone technology. It is just now being implemented. It is appearing in Europe first and the United States later. It uses an entirely different radio technology called orthogonal frequency division multiplexing (OFDM) to give even higher data rates up to 100 Mbps. More on that later.

How can you tell which of the above phone technologies you have? You cannot tell by the manufacturer, such as RIM BlackBerry, Samsung, LG, Motorola, Nokia, Sony Ericsson, and so on, because they all make multiple types. However, a few clues follow.

Carriers AT&T and T-Mobile provide GSM/WCDMA phones. Most also have some form of HSPA. These carriers will eventually morph their systems to LTE. Carriers Sprint and Verizon use cdma2000 phones. Most phones today are also what we call multimode phones in that they incorporate two or more different technologies. The most common combination is GSM/WCDMA/HSPA.

DIGITAL MODULATION AND WIRELESS TRANSMISSION METHODS

In Chapter 7, basic digital modulation methods ASK/OOK and FSK were covered. These are not used except for an advanced FSK as used in GSM. Instead, all cellular systems used advanced digital modulation methods. This section provides a look at these techniques.

Quadrature Amplitude Modulation

Quadrature amplitude modulation (QAM) is a technique that mixes both amplitude and phase variations in a carrier at the same time. What this technique does is to allow higher data rates within the same bandwidth. See the sidebar on spectral efficiency. The technique consists of transmitting multiple bits for each time interval of the carrier *symbol*. The term "symbol" means some unique combination of phase and amplitude. The term M as in M-QAM indicates how many bits are transmitted per time interval or symbol for each unique amplitude/phase combination.

The simplest form of QAM is 2-QAM, more commonly called QPSK or quadrature phase shift keying. It is produced by the circuit shown in Figure 8.5. It takes the serial bitstream and passes it through a 2-bit shift register producing two parallel bitstreams at half the rate. One is called the in-phase (I) bitstream and the other the quadrature (Q) bitstream. The I and Q signals are each applied to a mixer along with a local oscillator (LO) signal. The mixer output becomes part of the RF carrier. Note that the LO signal is applied directly to one mixer, but is shifted by 90 degrees and applied to the other mixer. The term "quadrature" means shifted by 90 degrees. The two mixer outputs are AM

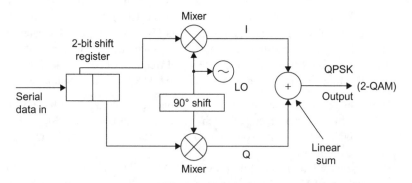

FIGURE 8.5 How QPSK is generated using mixers and a local oscillator shifted 90 degrees to produce in-phase (I) and quadrature (Q) signals that are summed.

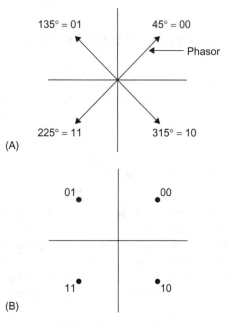

FIGURE 8.6 QPSK diagrams. **(A)** Phasor. **(B)** Constellation.

signals but at different phases. But when they are added together, the QPSK signal is generated. With this arrangement, you can transmit data at twice the rate of BPSK in the same bandwidth.

Figure 8.6 shows how we illustrate this with what we call the phasor and constellation diagrams. Figure 8.6A shows the amplitude and phase of each possible combination of 2 bits. The length of the arrow called a *phasor* is the amplitude, and its direction or angle is the phase. If the receiver gets a signal

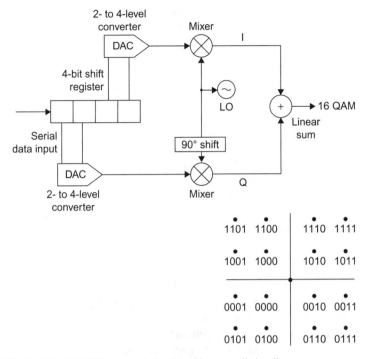

FIGURE 8.7 How 16-QAM is generated and resulting constellation diagram.

that is at a phase angle of 135 degrees, it means that the 2-bit combination 01 is being transmitted. The constellation diagram in Figure 8.6B is usually shown without the arrows or phasors.

This technique can be modified to produce 16-QAM, which transmits 4 bits per time interval or symbol causing the data rate to quadruple in the same bandwidth. The circuit for producing 16-QAM is shown in Figure 8.7. The 4-bit shift register produces two groups of I and Q bits. These are converted into four levels by a two- to four-level converter that works like a digital-to-analog converter (DAC). The resulting constellation diagram for 16-QAM is shown in Figure 8.6 as well. Other common versions are 64-QAM and 256-QAM, which transmit 6 and 8 bits per symbol, respectively, in the same bandwidth.

QPSK and QAM modulation methods are common in cable TV for digital TV as well as high-speed Internet service. They are also used with satellites and any broadband wireless application. All modern cell phones and most other digital wireless schemes use this type of modulation because of its spectral efficiency.

Orthogonal Frequency Division Multiplexing

Orthogonal frequency division multiplexing (OFDM) is a broadband modulation method like spread spectrum/CDMA. It takes a high-speed serial binary

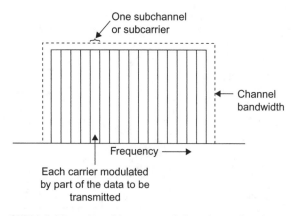

FIGURE 8.8 OFDM divides a channel into many subchannels or subcarriers, each of which is modulated by part of digital data to be transmitted.

signal and spreads it over a wide bandwidth. The serial data is passed through a circuit that maps out the constellation diagram for the modulation to be used. It then divides it into many slower-speed serial bitstreams. Each bitstream modulates a carrier on one of many adjacent carriers in the available bandwidth. This technique effectively divides a wide bandwidth into many narrower subchannels or subcarriers as shown in Figure 8.8. Sometimes dozens of channels are used, and in other cases hundreds or even thousands of carriers are used. The carrier frequencies are selected so they are orthogonal and as a result they will not interfere with one another even though they are directly adjacent. All the modulated channels are then added together and the combination transmitted in the available bandwidth. The type of modulation varies with the application but it is usually BPSK, QPSK, 16-QAM, or 64-QAM.

The basic technique is shown in Figure 8.9. The only practical way to create an OFDM signal is by using DSP. The DSP executes an algorithm called the inverse fast Fourier transform (IFFT) that creates the multiple adjacent modulated carriers. The DSP data is applied to DACs to create I and Q signals that are sent to mixers for creation of the final signal. At the receiver, the data is recovered by another DSP executing the FFT. The output is the original fast serial data.

OFDM seems impossibly complex while seemingly hogging spectrum space. Yet, it can transmit higher speeds in smaller bandwidths than most other types of digital modulation. It is very spectrally efficient. Furthermore, it is more resistant to multipath interference that can cause microwave links to lose data due to a reflected signal interfering with another or a direct signal causing cancellation and fading. Most of the new wireless technologies today use OFDM, including the forthcoming LTE cellular systems. Some examples of other OFDM uses include wireless local area networks (WLAN), wireless broadband (WiMAX), digital subscriber line (DSL) Internet access, digital TV, and AC power-line networking.

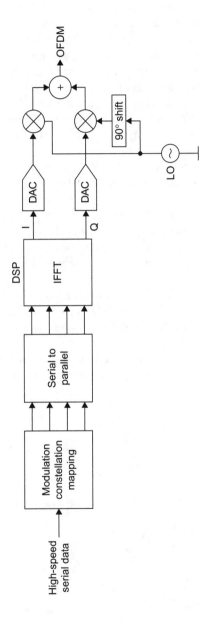

FIGURE 8.9   Generating OFDM is done by DSP and I/Q modulation.

Multiple Input Multiple Output

Another technique for boosting speed within a given bandwidth is multiple input multiple output (MIMO). Also known as spatial multiplexing, it is the use of multiple transmitters transmitting to multiple receivers, each sending different bitstreams on the same channel. The high-speed serial data is divided into two, three, four, or even more bitstreams, and each is used to modulate a different transmitter but operating on the same frequency. The signals from each transmitter reach the antennas of each receiver along with any reflected signals. See Figure 8.10, which shows four transmitters (Tx) and four receivers (Rx) or 4 × 4 MIMO. What allows the receivers to sort out each bitstream is the fact that each signal travels a different path, and will have different characteristics or spatial signature that can be sorted out by DSP in the receiver.

There are several things that make MIMO practical. First, MIMO is used at very high frequencies, especially microwave (>1 GHz). This means that antennas are short and inter-antenna space may be one wavelength or more, which is required for MIMO to work. The distance between the antennas ensures different spatial paths with different characteristics that let the receivers recover each bitstream. Second, thanks to semiconductor technology, it is possible to make the multiple transmitters and receivers very inexpensively, typically all on a single silicon chip.

MIMO has several benefits. First, it allows a higher data rate in the same channel bandwidth. Using a 4 × 4 MIMO, for example, would boost speed by a factor of 4. Configurations of MIMO can be any combination of transmitters and receivers. The 2 × 2 mode is popular but 4 × 4 is also used. Second, MIMO provides a more reliable link. It actually takes advantage of the multipath problems that usually plague microwave wireless and makes it

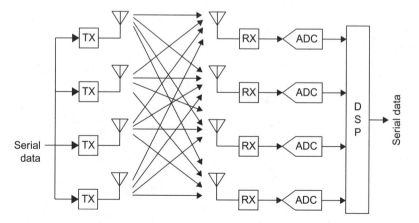

FIGURE 8.10   MIMO in a 4 × 4 configuration quadruples the data rate in an existing channel, while reducing the ill effects of multipath fading.

better. Finally, MIMO is usually combined with OFDM, which provides the ultimate data-rate boost, reliability, and spatial efficiency. MIMO is already used in wireless local area networks (WLAN) like Wi-Fi and in broadband wireless systems using WiMAX. Both of these technologies will be covered in the next chapter. MIMO will also be used with LTE, the next cell phone system. No, handsets probably will not have multiple antennas but the base stations will.

Spectral Efficiency

Spectral efficiency refers to how much data you can transmit in a given bandwidth. It is usually measured in bits per second per hertz (bps/Hz). ASK/OOK and FSK have efficiencies of <1. Special forms of FSK, such as Gaussian minimum shift keying (GMSK) used with GSM cell phones, are a bit better. BPSK has a spectral efficiency of 1. Table 8.1 gives the efficiencies of most of the popular digital wireless modulation schemes.

One last thing: the type of modulation is not the only thing that affects spectral efficiency. The level of noise in a channel also determines the number of bit errors that occur in transmission. The higher levels of QAM have more errors, so this limits their efficiency to some extent. Plain FSK, BPSK, and QPSK are very resistant to noise, so are more suitable in applications where noise is a problem. Noise and bit rates have to be traded off for good reliable transmission. And by the way, the spectral efficiency issue applies to wired as well as wireless communications channels.

TABLE 8.1 Spectral Efficiency of Selected Digital Wireless Modulation Schemes

Modulation Method	Spectral Efficiency (bps/Hz)
FSK	<1
GMSK	1.35
BPSK	1
QPSK	2
16-QAM	4
64-QAM	6
256-QAM	8
OFDM	Depends on modulation of subcarrier and number of subcarriers
MIMO	Depends on configuration

ACCESS METHODS AND DUPLEXING

Access methods are wireless techniques that allow multiple users to share a given part of the frequency spectrum. Access methods are also called *multiplexing*. Multiplexing means allowing two or more signals to be transmitted *concurrently* over the same communications medium such as a cable or within a given frequency band. Access methods allow many subscribers to share precious spectrum space. And they provide the means to implement full-duplex simultaneous transmit/receive that most phone users expect. There are four types: frequency division multiple access (FDMA), time division multiple access (TDMA), code division multiple access (CDMA), and orthogonal frequency division multiplex access (OFDMA). For full-duplex operation, there are two methods—frequency-division duplexing (FDD) and time-division duplexing (TDD).

FDMA—A given bandwidth is divided into many narrower channels as shown in Figure 8.11. A given radio will cover all of the bands but has the ability to select a specific operating frequency. GSM divides the band into 200-kHz channels. cdma2000 uses 1.25-MHz channels and WCDMA uses 5-MHz channels. The newer OFDM uses 5-, 10-, 15-, or 20-MHz channels.

FDD—Simultaneous transmit and receive is accomplished by using two different bands where half are used for transmit and the other half for receive. The transmitter and receiver operate at the same time but on different frequencies so they do not interfere with one another (see Figure 8.11).

TDMA—With this method, a single-frequency channel is used, but the signal is divided into fixed-duration time slots into which digitized segments of voice or data are transmitted. In digital voice transmission, the analog voice

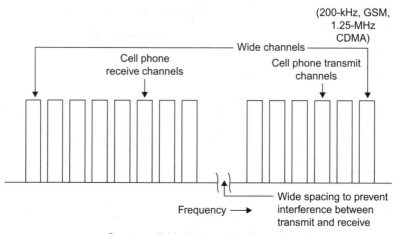

Spectrum divided into smaller channels

FIGURE 8.11 Frequency-division multiple access and frequency duplexing use two separate bands of frequencies divided into individual channels.

signal is periodically sampled and digitized into a stream of binary numbers. These binary values are then compressed into a smaller number of bits by a vocoder circuit. The resulting bits are then transmitted to the receiver. With TDMA, binary values from two or more voice sources can be interleaved and sequenced, and then sent over the single channel. For example, in GSM systems, there are eight time slots, as shown in Figure 8.12. The data is transmitted at a 270-kbps rate in a 200-kHz wide channel. Each time slot contains 1 byte of the voice data for one phone call. The frame of eight signals is then repeated.

TDD—Time-division duplexing facilitates concurrent send and receive by assigning transmitted signals in one time slot and received signals in another time slot. They share the same frequency channel. Because all this is happening at a high rate of speed, you cannot determine that transmit and receive are not simultaneous, although in reality they are not.

CDMA—This method uses spread-spectrum technology. It takes the digitized voice and further processes it to give it special coding, and then spreads it over a wide bandwidth. In cdma2000 phones, the channel bandwidth is 1.25 MHz wide, and all cell phone signals are coded and spread over this bandwidth. The process goes like this: the analog voice signal produced by the microphone is digitized in an ADC, and then compressed by a special digital circuit called a *vocoder* or *voice coder*. This reduces the total number of bits and the speed of the digital signal. This vocoded signal is then mixed with a high-speed serial code signal, making it unique so that the receiver will recognize it. This is done with an XOR gate as shown in Figure 8.13. The higher-frequency signal, at 1.23 Mbps, is called the *chipping data*. The resulting signal fills the 1.25-MHz channel. In WCDMA systems, the process is the same, but the chipping signal is 3.84 Mbps and the channel width is 5 MHz.

Many cell phone calls are then transmitted on the same channel. They actually all mix together, but because they occur randomly with one another, they do not interfere with one another. Typically, up to about 64 users can share the same bandwidth. One signal simply appears as a low-level noise to another. The desired signal is picked up at the receiver by recognizing the unique code given to the transmitted signal. The two most common types of spread spectrum are direct

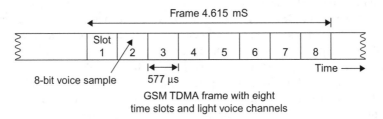

GSM TDMA frame with eight
time slots and light voice channels

FIGURE 8.12   TDMA multiplexing divides a channel into sequential time slots for the digital data.

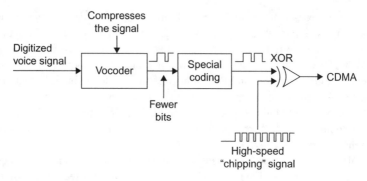

FIGURE 8.13   How CDMA is generated.

sequence and frequency hopping. Direct-sequence spread spectrum (DSSS) is used in cell phones as illustrated in Figure 8.13. Frequency hopping is used in Bluetooth wireless systems (see Chapter 9). Most duplexing is by FDD, but some of the newer versions of WCDMA also use TDD.

OFDMA

Access with OFDM is done by just assigning one or more of the subchannels to each phone call. With each subchannel 15 kHz wide, each is capable of carrying a compressed (vocoded) voice signal so that many calls can be handled. In practice, more than one channel may be used, depending on the specific frequency assignments, bandwidths, and other features of the LTE system being used.

A LOOK INSIDE A 3G CELL PHONE

Figure 8.14 shows the insides of a modern 3G cell phone. This one uses the WCDMA/HSPA 3G technology, but is also backward compatible with GSM/ EDGE if an area does not have 3G yet. There is a separate radio transceiver (XCVR) for each. All of the baseband functions like modulation and demodulation and handset management are handled by one or more processors. One processor may be DSP. This handset also has a Bluetooth radio for the headset (not shown) and a Wi-Fi transceiver for Internet access via a wireless LAN access point or hot spot. These technologies will be covered in the next chapter. A GPS radio provides navigation as well as fulfills the E911 requirement. Note that all radios have separate antennas. A camera in included. The CCD is the charge-coupled device that converts the light and color variations in the scene into electrical signals. The stereo headphones are for the MP3 music player.

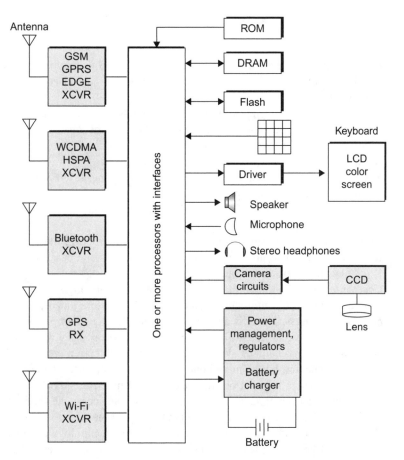

FIGURE 8.14   A block diagram of 3G cell phone with multiple radios and functions.

THE LATEST CELL PHONE TECHNOLOGY

The cell phone industry is one of the fastest changing in electronics. New products come out almost daily. And new features and capabilities are invented regularly to keep the technology fresh and the sales continuous. Here is a snapshot in time of the latest cell phone developments.

Smartphones

A smartphone is a basic cell phone but with advanced features that expand its use. The basic cell phone does phone calls or at most handles text messaging as well. But a smartphone has multiple functions. Some examples of smartphones are the BlackBerry models from Research in Motion (RIM), Apple's iPhone, some Nokia models, the Palm Pre, and various models from Samsung,

LG, and Motorola. They are too numerous to mention and they go out of date monthly or yearly.

High-speed data capability. Most smartphones use 3G technology, so they can deliver almost as fast a data rate as your home Internet connections, usually 1 to 2 Mbps or better.

Internet access. A built-in browser lets you access the Internet just as you would from your home laptop. The main limitation is the small screen size, which prevents you from doing as much as you would like, but you can still do it.

Email. Email is one of the key features of a smartphone. It lets you stay in touch without a PC or laptop.

Messaging. It goes without saying that text messaging is a main data feature.

Digital camera. Most smartphones have an excellent high-resolution camera built in. It is not as great as a stand-alone digital camera, but it does have 1 to 3 megapixels of resolution.

Camcorder. Some of the digital cameras have video capability as well. The main limitation is the amount of memory available to store videos.

MP3/iPod. A music player is a standard feature of many smart phones. The key here is to have sufficient flash memory to store the songs you want.

Wireless local area network (WLAN) access. You will learn more about this in the next chapter, but a smartphone has a built-in separate transceiver to connect to a Wi-Fi hot spot or other access point in airports, hotels, coffee shops, and other locations just like a laptop. You can access the Internet or email as an alternative to doing it through the cellular network.

Bluetooth. Bluetooth is another short-range wireless technology covered in Chapter 9. It is used in smartphones for wireless headsets. A headset contains a headphone and a microphone that fits in the ear. It talks to the cell phone wirelessly via a Bluetooth link.

GPS. The global positioning system is the worldwide satellite navigation system. Many phones have a built-in GPS receiver so that you can find your way. Some even offer full navigation software and display as in those personal navigation devices (PNDs) like those from Garmin and Tom Tom.

Femtocells

One of the most frustrating problems cell phone users have is poor coverage at home. Many homes are not located near cell sites or have poor indoor performance. This is maddening to many, especially those who want to give up their regular wired home phone and go only wireless. One solution to this problem is the femtocell. This is a small base station designed to be installed at home. The concept is illustrated in Figure 8.15. It can usually accommodate calls from up to four cell phones. The femtocell connects back to the carrier via your high-speed Internet connection, a DSL line, or a cable TV link.

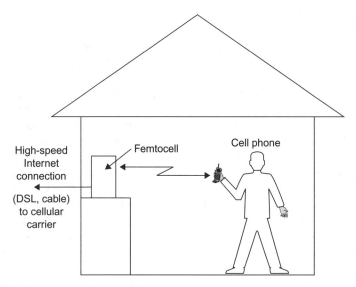

FIGURE 8.15   Concept of femtocell.

It provides great in-home coverage and has the added benefit that it offloads to the cell site and the carrier, providing more capacity to others.

Femtocells are just now becoming available but will soon be rolled out nationally. It is expected that the new LTE systems will make extensive use of femtocells.

Mobile TV

Video is now available on some phones. This video comes to you over the cell phone network, so it uses the high-speed data connection. Video is very data intensive, so it really loads down the network, limits subscriber capacity, and slows speed in some cases and raises your monthly bill. Yet, you can get various TV programs plus news, weather, sports, and comedy this way.

It has been decided that the best way to offer TV on a cell phone is to broadcast it directly to a small receiver in the cell phone. This keeps the video data overload off the cellular network and provides a better picture and more programming choices. This over the air (OTA) cell phone video is now available in many U.S. cities with more to come. A company and service called MediaFLO is transmitting TV to those who subscribe. Up to 15 or 20 channels provide a mix of programming. Some TV is also available in Europe using an OFDM system called Digital Video Broadcast-Handset (DVB-H). Variations are also available in Japan and Korea.

It is expected that some time in the near future, the U.S. digital TV system called the Advanced Television Systems Committee (ATSC) standard will be available in a mobile version that can be received via free OTA if the phone has a receiver built in. In all systems, the video is highly compressed using a standard referred to as H.264 and reformatted for a 360 $\times$ 240 pixel screen.

Location Technology

Location technology refers to systems and circuits that are designed to pinpoint the location of a cell phone in use. This can be done now to a certain extent, as a carrier can always determine which cell site a phone is currently using. That only locates the phone within an area of a few miles. The FCC has mandated that all cell phones can be located within 50 to 100 feet in case of an emergency. This is part of the E911 system for all cell phones. E911 means that your cell phone company has a system that will send your physical location to emergency services if you call 911 from your cell phone. Some systems use GPS, but others have another type of location technology. The whole thing is transparent to a user, and most users actually do not know that their phone has this capability.

The location capability has many concerned for privacy reasons. But if you don't want to be located, turn off your phone. Otherwise, learn to live with it, as it many save your life one day. Location technology is also supposed to lead to services that are based on location, such as advertisements for restaurants and other businesses near you.

When GPS is used, the GPS receiver computes your coordinates from the satellite signals and sends them to the carrier for use if called for. In another system, your location is determined by actually triangulating your phone with three nearby cell sites. No additional electronics are needed inside the phone, but special antennas, receivers, and other gear are needed at the cell sites. Both systems can usually pinpoint you within 50 feet or so depending on terrain and environment.

Project 8.1

Examine a Base Station

Go look at a cell site. The wireless phone company won't let you inside, but you can drive close to many and examine what you can see. Especially note the antenna arrangement at the top of the tower. Most have a triangular arrangement of three antenna arrays that divide the coverage into three 120-degree wide areas. You will see multiple coax cables going up the tower to carry the received and transmitted signals. Check for a dish that connects the site to the main switching station.

Project 8.2

Check Out the Latest Handsets, Features, and Services

Go to a cell phone store and ask about what digital services are available in your area. Try to find out what the technology is, such as GPRS, EDGE, cdma2000, 1xRTT, WCDMA, HSPA, and so on. Do you have email, instant messaging, games, Internet access, transmission of color digital photos, or what?

Project 8.3

Get an Area Coverage Map

Call your cellular carrier and ask for a coverage map. Some carriers actually give you this, but many times you have to ask for it. It may also be on your carrier website. The map will show you the actual range and coverage of the system. If you leave the area, you may have no coverage or you may switch over to another system in a roam mode.

Project 8.4

ID the Carriers

Identify the cell carriers (ATT Wireless, Cingular, Verizon, Sprint, Alltel, MetroPCS, etc.) in your area. If they advertise locally in the newspapers or in cell phone stores, they cover your area.

Project 8.5

Relevant Websites

There are several great websites where you can get additional information on how cell phones work. Here are a couple that I like: www.howstuffworks.com and www.sss-mag.com.

Chapter 9

Networking: Wired and Wireless

All Devices Talking to One Another

> **In this Chapter:**
> - Types of networks defined: WAN, MAN, LAN, and PAN.
> - Ethernet, Sonet, and other wired networking technologies.
> - Internet transmission fundamentals.
> - Wireless networking technologies, including Wi-Fi, Bluetooth, WiMAX, ZigBee, and others.

INTRODUCTION

You already know from reading this book that virtually every electronic product has a digital computer built into it in the form of an embedded controller or microcomputer. But one other fact about electronic equipment that you may not know is that a large percentage of these devices are now being interconnected to one another to form networks. This is particularly true of computers. After all, what computer today is not connected to some form of network? And while we may be nearing the point where all computers are networked, the trend continues toward networking many other non-computer devices. This chapter takes a look at networking concepts and the most popular forms of networks being used today, both wired and wireless.

IS EVERYTHING NETWORKED?

A network is just a system of people or things that are connected to one another. If someone was to ask you to give an example of a network, you probably would use the computer network definition. But there are all sorts of other networks in which we all participate in some form or another. One of the biggest networks is our interconnecting road system. Our highways and byways interconnect cities with one another and provide transportation routes for people and goods. The railroad is a massive network. You could also call the airline system a massive network of cities connected by airplane flights. Our whole electrical grid is a massive network. And one of the largest networks in

doi: 10.1016/B978-1-85617-700-9.00009-6

the world is the telephone system, both wired and wireless. And then again, let's not forget the Internet itself which connects us all to everything else. For our purposes here, our network definition is a system designed for communications via computers talking to one another or individuals talking to one another over cell phones. In all cases, a network is involved.

TYPES OF NETWORKS

There are lots of different ways to interconnect individual computers or other electronic products. You can do it by electrical cable or fiber optic cable or you can do it in a wireless manner. In any case, each computer or user in the network is called a *node*. The goal of a network is to connect each node to all other nodes in some way so that they can communicate. Over the years, several different levels of networks have evolved.

Wide Area Networks

A wide area network (WAN) is the largest network of all and we could consider the entire telephone network or the Internet as a large WAN. Sometimes WANs are smaller and may be localized to a country, a state, a city, or other large geographical area. The Internet is certainly a WAN. Most WANs are made up of fiber optic cable so that they can carry lots of data at very high speeds.

Metropolitan Area Networks

A metropolitan area network (MAN) is a network that covers a smaller geographical area such as a city or a large college campus system. Good examples of MANs are the local telephone company and your local cable TV company. MANs are also widely used in large companies and in governments to interconnect their computers. Most MANs are also implemented with fiber optic cable to maintain speed and data capacity.

Local Area Networks

A local area network (LAN) covers a small group of computers, typically a thousand computers or less. LANs are implemented in companies, small businesses, and even in homes. A LAN may only have a couple of users but could have several hundred, depending on the circumstances. Even two connected PCs at home comprise a LAN. In most cases, LANs are connected externally to a MAN or a WAN for external communications.

Personal Area Networks

Most personal area networks (PANs) are short-range wireless networks. A PAN is formed when two or more computers or cell phones interconnect to one another wirelessly over a short range, typically less than about 30 feet. Most PANs are what we call ad hoc networks that are only set up temporarily for a specific purpose. The Bluetooth headset option on a cell phone is an example of a PAN.

Home Area Networks

A home area network (HAN) is a network inside the home used to provide monitoring and control over energy usage. It connects to the utility's electric meter and monitors energy usage so the home owner can see what energy is being used and where. It is also used to give the utility a way to control the heating and air conditioning to save energy. The HAN also provides a way to conveniently turn appliances off and on. The HAN may use wireless or communications over the AC power line. The HAN is the home portion of the national Smart Grid effort to conserve energy.

Storage Area Networks

A storage area network (SAN), as its name implies, is a network that interconnects large data storage devices to computers and to one another. Large arrays of hard disks are usually set up to store massive amounts of data needed by business, the government, and other organizations. The SAN provides a way to access all this information in a fast and easy way.

Network Relationships

As it turns out, the different types of networks are connected to one another, the result being one massive network. Figure 9.1 is an example. For example, local

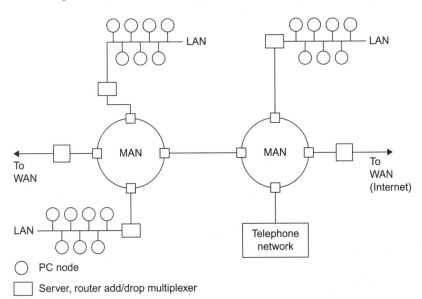

FIGURE 9.1 How LANs, MANs, and WANs connect to form a huge worldwide network of computers and other devices.

LANs are connected to MANs, while the MANs themselves are connected to WANs, and so on. In this figure, the circles represent PCs while the squares represent network connecting points. These include servers, the computers that manage LANs and MANs; routers, which determine connections in the network; and add/drop multiplexers that put data on the network and take it off. All the interconnecting links are fiber optic cable except for the copper cable in the LANs. This complex hierarchy essentially lets any computer talk to any other computer under the right conditions. The Internet implements such a hierarchy of networks.

NETWORK INTERCONNECTION METHODS

As you can imagine, there are lots of different ways to connect computers or other electronic devices to one another. For example, you could run a cable between each of the computers you want to interconnect. The result would be a large, complex, and very messy and expensive interconnection system. It is called a *mesh*. Except for a few computers, such systems are not used. Instead, over the years, several different interconnection methods called network topologies have emerged. These are the star, the ring, and the bus, all simply illustrated in Figure 9.2.

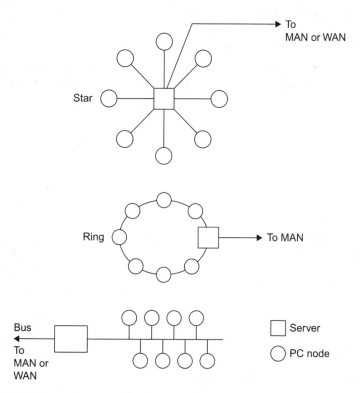

FIGURE 9.2   Three main networking topologies—star, ring, and bus. Bus is the most widely used.

In the *star* method, each node is connected to a central control computer or server that manages the interconnections from any PC to any other. In the *ring* interconnection, the computers are simply connected into a single closed loop. Then data is transferred from one computer to the next by sending it around the ring. The computer destined for the information will recognize it and grab it off the ring as it comes by. The *bus* method of interconnection is one of the most popular, as it is simply a common cable or connecting point to which all computers are attached. The problem with the bus is that there can only be one sender or transmitting party of the information on the bus at a time. This requires some method of managing who transmits to whom and when. All nodes can receive.

The mesh network mentioned earlier is a messy thing to implement with cables. However, it can be implemented wirelessly. In a mesh, all nodes can talk to all other nodes within their range. This is advantageous in that there are multiple paths for data to take if one path should be blocked for some reason. More on that later.

WIRED NETWORKING TECHNOLOGIES

Wired networks use three types of cable: twisted pair, coax, or fiber optic. The telephone system has used twisted-pair cable from the very beginning and it is still in use today. It is also the predominant way that LANs are interconnected. Coax cable was once popular but is not widely used today except in cable TV systems. Fiber optic cables were not practical until the 1970s, but since then have been developed into the best way to wire computer networks because of their high speed and low loss.

Types of Cables

Coax cable was the original networking medium (see Figure 9.3). It has a copper wire center conductor surrounded by an insulator like Teflon or another type of plastic, which in turn is usually covered with a wire-mesh braid.

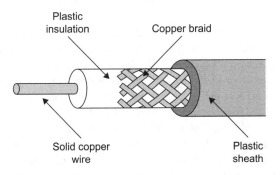

FIGURE 9.3   Coax cable. It is now mainly used in cable TV networks. The most common is RG-6/U and uses F-type connectors.

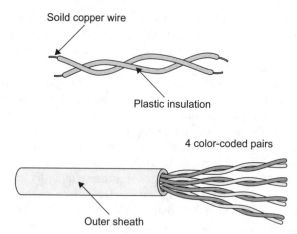

Soild copper wire

Plastic insulation

4 color-coded pairs

Outer sheath

FIGURE 9.4   Twisted-pair cable. Initially used for telephones and is still used. Computer net-work cables use four twisted pairs. A shielded version is available where noise is a problem.

Category	Maximum Data Rate	Application/Comment
TABLE 9.1 Types of Twisted-Pair Cable		
CAT5	100 Mbps	Ethernet LAN, telephones
CAT5e	1 Gbps	Ethernet LAN
CAT6	10 Gbps	Ethernet LAN, video
CAT7	10 Gbps	LAN, point to point

A plastic outer jacket covers the whole thing. The center wire and the mesh are the two connection points. Coax can handle frequencies up to many gigahertz. But it has high losses over long distances. Coax is mainly used in cable TV networks.

Twisted pair is the widely used cable for telephones and LANs. It consists of two solid insulated wires twisted together as shown in Figure 9.4. Most telephone or LAN cables contain multiple pairs. Typical LAN cables have four pairs in a single cable. These cables are standardized and categorized as shown in Table 9.1. CAT5 and CAT5e are the most common. The eight wires are usually terminated in a connector called an RJ-45.

Fiber optic cable consists of a thin glass or plastic center piece surrounded by a plastic outer cover. Then that is further wrapped by an outer plastic cover-ing (see Figure 9.5). Light waves are passed down the center glass or plastic cable. Most light comes from an infrared (IR) light laser. You cannot see IR but it is light nonetheless. IR light is what your TV remote control uses. The IR light is pulsed off and on to create binary signals. These are picked up

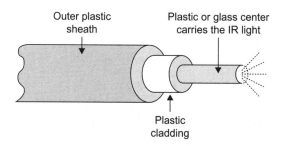

FIGURE 9.5   Fiber optic cable. It is nonconductive because it is made of plastic and glass, and is used to transmit data as infrared (IR) light pulses.

at the receiving end by a light detector and converted back into binary 0 and 1 voltages (see Figure 9.6).

Over the years, dozens of different wired networks have been developed. But in the past two decades, only two types of wired networks are dominant: Ethernet and Sonet.

Ethernet

Ethernet was originally developed as a LAN in the 1970s. It used coax cable in a bus topology to interconnect multiple computers. Ethernet is also known by its Institute of Electrical and Electronic Engineers (IEEE) standard designation 802.3. There are many variations of this technology, but the most common one uses unshielded twisted-pair cable. The bus topology is generally implemented inside a piece of equipment called a *hub* or *switch* (see Figure 9.7). The cable from each computer or other device is connected to one of the jacks on the hub or switch. The hub or switch connects to a master computer called a *server* that manages the network. Generally speaking, connections to the hub or switch are limited in length to about 100 meters or 300 feet.

The initial Ethernet LAN had a maximum data rate of 2.93 Mbps over coax. When Ethernet was standardized, the basic data rate was 10 Mbps. Over the years, Ethernet has been improved and developed to create versions with data speeds over twisted-pair cable of 100 Mbps, 1 Gbps, and 10 Gbps. While 10 Mbps can be implemented over a single twisted pair, to achieve speeds of 100 Mbps, 1 Gbps, and 10 Gbps, all four twisted pairs in the CAT5 cable are used. The data is divided between and transmitted in parallel at the same time using special multilevel encoding that permits such higher data rates to be achieved.

There are also versions of Ethernet that use fiber optic cable. Fiber optic cable is normally used on the 1-Gbps and 10-Gbps versions. The latest versions of Ethernet provide 40-Gbps and 100-Gbps data rates.

How Is Ethernet Used?

Ethernet uses a bus topology connection where only one node can transmit at a time. This requires the use of an access method that permits the nodes on

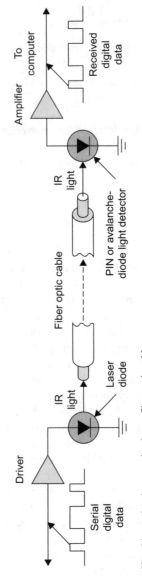

FIGURE 9.6 How binary data is transmitted over fiber optic cable.

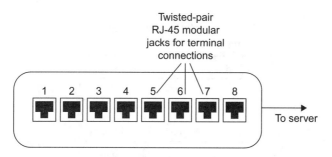

FIGURE 9.7 Ethernet hub or switch implements a bus that is accessed by RJ-45 connectors and twisted-pair cable.

Preamble	Start of frame delimiter	Destination address	Source address	Length	Data	Frame check sequence
7 bytes	1 byte	6 bytes	6 bytes	2 bytes	46–1500 bytes	4 bytes

FIGURE 9.8 Ethernet frame or packet.

the bus to share it. Ethernet uses a technique known as *carrier sense multiple access with collision detection* (CSMA/CD). This is part of the IEEE 802.3 standard. Essentially what happens is that if a computer wants to transmit, it monitors the bus, and if no one is transmitting, it will begin transmitting its signal. If someone else tries to transmit at the same time, a collision occurs and both computers stop transmitting. Each will wait a random period of time and then try again. Whoever captures the bus first will be allowed to transmit. The data is transmitted over the bus and any computer connected to it can receive it. However, typically the transmission is only intended for one other computer. To identify that computer, each is given an Ethernet address that is transmitted along with the data to be sent.

The data to be transmitted are packaged together in a group of bits called a packet. Figure 9.8 shows what a typical Ethernet packet looks like. The preamble consists of 7 bytes of alternating 1's and 0's that help the receiving computer establish clock synchronization with the transmitting computer. The start frame delimiter presents a unique 1-byte code indicating that a frame of Ethernet data is to be transmitted. Next, a 6-byte destination address identifies the desired receiving node. The next 6 bytes identify the sending node. This is followed by another 2 bytes of data that indicate how long the data message is. The block of data is then transmitted. It can be any length from about 46 to 1500 bytes. The packet ends with a frame check sequence that is essentially a unique code designed for error detection and correction. If bit errors occur during transmission, the receiving computer can perform specific logic functions that allow the error to be detected and then corrected.

TABLE 9.2 Sonet/SDH Designations and Data Rates	
Sonet Level	Data Rate
OC-1	51.84 Mbps
OC-3	155.52 Mbps
OC-12	622.08 Mbps
OC-48	2.488 Gbps
OC-192	9.953 Gbps
OC-768	39.812 Gbps

While Ethernet started out as a LAN, it has morphed into other larger forms of networks. Because of its high-speed capability with fiber optic cable, it is also used to form MANs and even WANs. A special version for MANs and WANs is called *Carrier Ethernet*. There is also a version of Ethernet used in storage area networks, called *Internet Small Systems Computer Interface* or *iSCSI*. There is also a wireless version of Ethernet that will be discussed in another section in this chapter.

Sonet

Sonet means *synchronous optical network*. Also known by its international standard name, *synchronous digital hierarchy*, this fiber optic network system was designed specifically for high-speed data transmission in telephone networks and computer systems. Basically, Sonet defines how data is transmitted over the fiber optic cables. This includes the data format as well as data speeds. Table 9.2 shows the various standardized transmission speeds from 51.84 to 39.812 Gbps. The bulk of Sonet installations, which are used primarily in the telephone networks and the Internet, use 2.488 Gbps (also generally called 2.5 Gbps) and 9.953 Gbps (generally referred to as 10-Gbps Sonet). More recently, new 39.812 (40) Gbps networks are being built for the Internet. Data is transmitted in fixed-length frames made up of 810 bytes, most of which is data. The network topology is either point to point or i n a ring.

Most big WANs and MANs use the Sonet system. However, Sonet is often used to transmit other data, including the telephone company T1 and T3 line data, asynchronous transfer mode (ATM), frame relay, and other telephone company formats. It is also widely used to carry Ethernet packets as well as the unique packets that are part of the Internet protocol.

Telephone Data Systems

When the wired telephone system converted to digital many years ago, several key data transmission standards were developed. The most common are the T carrier systems, as well as frame relay and asynchronous transfer mode (ATM). The most widely used digital telephone networking standard is the T-1 system. It was developed to transmit 24 digitized telephone lines over a single cable. The voice signals were digitized into 8-bit chunks and then transmitted sequentially over a T-1 line from one central office to another. The basic T-1 frame format is shown in Figure 9.9. It consists of 192 data bits made up of 24 bytes representing individual chunks of telephone calls. An additional synchronizing bit is added. The result is a T-1 signal that transmits at a rate of 1.544 Mbps. And while T-1 is still widely used for digital telephone calls, it is also used for broadband computer data connections and cell phone backhaul.

There are other T carrier systems that are used to multiplex T-1 lines onto larger cables. The most common is the T-3, which allows dozens of T-1 lines to be multiplexed on a single cable at a 44.736-Mbps data rate.

Frame relay transmits data mostly in telephone systems. Both the T carrier and frame relay are synchronous systems, meaning that they are synchronized to a master clock signal.

Asynchronous transfer mode (ATM), as its name implies, is an asynchronous system that transmits data in packets. Each packet is made up of 53 bytes—5 bytes for a header that designates the destination and a 48-byte payload segment. Packets may be digital voice or computer data.

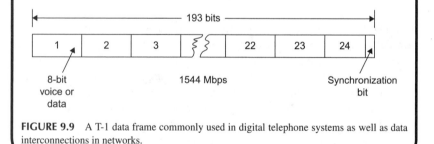

FIGURE 9.9 A T-1 data frame commonly used in digital telephone systems as well as data interconnections in networks.

Optical Transport Network

The optical transport network (OTN) is a technology used to implement the Internet backbone network. This is the core long haul fiber optical network that connects the world together. OTN uses special polarization and modulation techniques to transmit data rates of 40-Gbps and 100-Gbps.

HOW THE INTERNET WORKS

The Internet is an enormous WAN covering the entire world. It consists of many super-fast, fiber optic cable links that are generally referred to as the backbone of

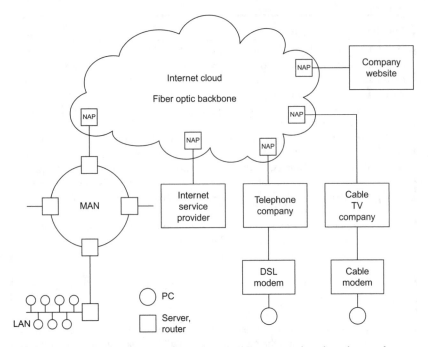

FIGURE 9.10 Internet illustrated as cloud of fiber backbone connections through network access points (NAP) that attach to WANs, MANs, and LANs.

the network. These high-speed links are owned by companies and governments and provide connection points for companies, governments, WANs, Internet service providers, and hosts for websites. These connection points are known as network access points (NAPs). See Figure 9.10. There are dozens of NAPs in the United States and they provide the link to the backbone connections. These backbone connections are worldwide. Because they are complex and interlinked, they are often referred to as the Internet cloud. Since data being transmitted from one point to the other can go by multiple paths in the backbone, there is no way to know the exact path, so the cloud designation is appropriate. You put the data transmission into one NAP entry point to the cloud and it comes out in another. From there it goes to other MANs and LANs for connections to other networks and to individual users.

Data is transmitted over the Internet using a standardized method referred to as TCP/IP. This means transmission control protocol/Internet protocol. Both are protocols for building packets that are transmitted over the Internet.

Assume that a LAN and MAN using Ethernet want to transmit information over the Internet. The Ethernet packets are then sent to a NAP where they are assembled into the packets that actually get transmitted over the fiber optic cable. The TCP assembles the data to be transmitted into unique packets and attaches a header to it, giving the source and destination addresses, a sequence

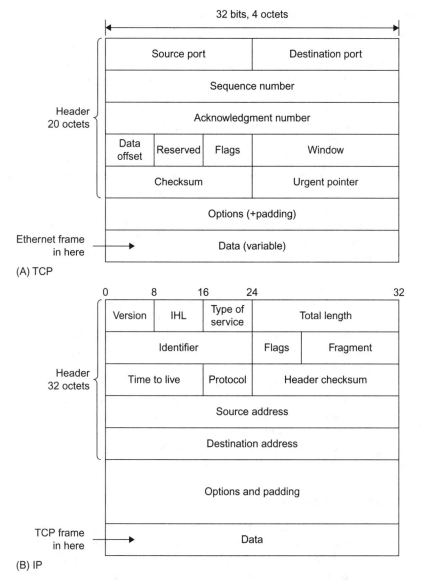

FIGURE 9.11 Protocols used in Internet. **(A)** Transmission control protocol (TCP). **(B)** Internet protocol (IP). The data in each frame is transmitted from left to right and top to bottom.

number, and some error detection and correction information. The general format is shown in Figure 9.11A.

The TCP packets then become the data that is assembled further into Internet packets (see Figure 9.11B). The Internet protocol attaches a header giving source and destination addresses as well as other information about where and how the data is to be sent.

The IP header contains a 32-bit destination address identifying the location of the computer receiving the data. This 32-bit address is divided into four 8-bit segments. Each 8-bit segment is usually referred to by its binary equivalent. The result is the IP address. Every computer and most other devices have an IP address assigned by an organization set up for assigning and managing Internet addresses. You may have seen the address referred to in the format 51.78.23.189. This is called the *dotted decimal format*.

With a 32-bit address, over 4 billion individual computers and other devices can be assigned a unique address. However, we are already beginning to run out of addresses and a new updated form of Internet protocol called IPv6 (Internet Protocol version 6) is gradually replacing it. It has a 128-bit address that provides billions and trillions of additional addresses so that even the smallest inconsequential device can have its own Internet address.

As the data is transmitted over the Internet, it passes through many devices referred to as *routers*. The router is the basic transmission control device of the Internet. What the router does is examine all of the packets transmitted to it and specifically looks at their IP destination addresses. The router temporarily will store data being transmitted and then decide which part of the network it should be connected to next to reach its desired destination. Routers store sets of information called *routing tables* that help the router decide the best transmission path through the Internet.

The data being transmitted ultimately reaches the NAP connected to the network where the destination exists. This may be a company, MAN, or LAN, or it could be an Internet service provider, such as the telephone company or cable TV company that will ultimately connect to the source computer. The source computer then strips away the TCP/IP headers to recover the original Sonet or Ethernet data.

WIRELESS NETWORKS

Wireless networks are usually LANs or PANs, but may also be MANs. Most wireless networks are short-range wireless connections that bring greater mobility as well as freedom from interconnecting cables to networking. Figure 9.12 shows a graph designating the various types of wireless networking schemes in use today. In this graph, each networking technology is defined in terms of its data rate as well as its range. In this section, we take a look at all of these technologies, most of which you have heard of or already use on a regular basis.

Wi-Fi

The most widely used wireless networking technology is referred to by its trade name Wi-Fi, meaning *wireless fidelity*. This the name given to the IEEE's

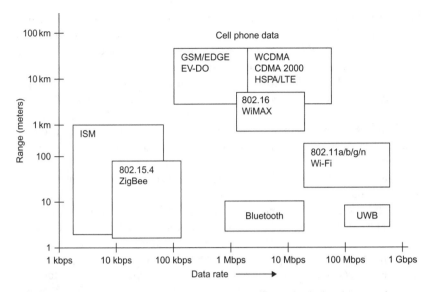

FIGURE 9.12 Summary of short-range wireless networking technologies shown as data rate versus range.

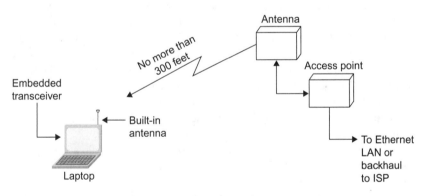

FIGURE 9.13 How a Wi-Fi access point or hot spot works.

wireless Ethernet standard designated as 802.11. Wi-Fi was originally developed as a wireless extension to normal Ethernet LANs. It allowed laptops and personal computers and even cell phones to connect to a company or organizational LAN wirelessly instead of the normal CAT5 connection. This not only simplified and in many case eliminated new wiring but also gave employees greater freedom and mobility, particularly since most employees use laptop computers rather than fixed PCs.

Figure 9.13 shows how Wi-Fi works. Each PC, laptop, or cell phone has a built-in data transceiver and antenna that communicate with a wireless access point (AP). This wireless access point is a transceiver that connects directly to an Ethernet switch, which, in turn, connects to the LAN wiring and its server. Access points are placed at strategic locations within buildings to allow as many people as possible to connect wirelessly.

There are also public access points known as *hot spots*. These are present in coffee shops, airports, hotels, and in many other locations, allowing anyone with a laptop or cell phone to connect. These hot spots talk back to a central point by a backhaul method, such as a T1 telephone line or microwave relay link that connects to an Internet service provider that, in turn, is connected to the Internet.

The 802.11 standard exists in several basic forms. The most popular version of this standard is 802.11b. It operates in the 2.4- to 2.483-GHz unlicensed band. It uses the direct sequence spread spectrum (DSSS) and is capable of data rates up to 11 Mbps in a range of up to 100 meters. This frequency band is divided into 11 channels, each 22 MHz wide. The computer and access point decide on which channel to use to minimize interference with other transmissions. If you are close enough to the access point, you can actually get the 11-Mbps data rate, but if you are not, the system automatically ratchets down to 5.5, 2, or even 1 bps depending on the range, interference, and noise or other environmental conditions.

Another popular version of the standard is 802.11g. It is a faster version of Wi-Fi that also operates in the 2.4-GHz band. It can achieve data rates as high as 54 Mbps up to 100 meters. It uses OFDM with 52 subcarriers. The most recent standard is 802.11n, which uses OFDM and MIMO to achieve even higher data rates up to approximately 600 Mbps. Another less popular version is 802.11a, which uses the 5-GHz unlicensed band. It too uses OFDM and can achieve data rates to 54 Mbps. The IEEE is also currently working on an advanced wireless standard that should ultimately be able to achieve a 1-Gbps data rate in the near future.

Wi-Fi is also built into most of the new smartphones. While a smartphone has data capability because of its 3G digital technology, Wi-Fi gives it one more way to access the Internet and email.

Wi-Fi is also the most popular home networking technology. With no wires to install, it is easy to set up and use with a broadband DSL connection or cable TV access.

Bluetooth

Bluetooth was developed as a personal area network (PAN) for connecting computers, cell phones, and other devices up to a range of about 30 feet. It is a low-speed data transmission method. Bluetooth operates in the same 2.4- to 2.483-GHz unlicensed spectrum as Wi-Fi. It uses a technique known as frequency hopping spread spectrum (FHSS), where the data is divided into chunks and transmitted via a carrier that hops from one random frequency to another. Data is transmitted at a 1-Mbps rate using FSK. An enhanced data rate (EDR) version of Bluetooth is also available to transmit at higher speeds up to 3 Mbps. Even newer versions can give data rates as high as 24 Mbps and even higher rates are possible in future versions of the standard. The standard is managed by the Bluetooth Special Interest Group (SIG).

One of the basic features of Bluetooth is that it is capable of forming small networks called piconets. It does this by linking two Bluetooth devices together. One serves as a master controller and it can connect to seven other Bluetooth slave devices. Once the PAN has been set up, the various connected devices can exchange information with one another through the master.

By far the most common application for Bluetooth is cordless headsets for cell phones. But you will also encounter it in some wireless connections between laptops and cell phones. It is also used in some computer peripheral devices such as PC-to-printer connections.

ZigBee

ZigBee is another short-range PAN network technology with the IEEE designation 802.15.4. It uses low power, so the range is typically 100 meters or less, basically depending on the antennas and the physical environment where it is used. ZigBee also operates in that unlicensed spectrum from 2.4 to 2.483 GHz.

The IEEE 802.15.4 standard defines the basic radio technology. It uses direct-frequency spread spectrum (DSSS) and a version of QPSK that gives a data rate of 250 Kbps. The frequency spectrum is divided into 16 1-MHz channels.

The basic function for ZigBee is monitoring and control. Monitoring refers to telemetry that allows sensors to collect information, which is then transmitted by radio to some central collection point. It also allows for remote control of the devices such as lights, motors, and other items. The ZigBee Alliance, the organization that promotes and enhances this wireless technology, has developed a series of applications for industrial monitoring and control, home monitoring and control, energy monitoring (automatic meter reading) and control, and several others.

A major feature of the ZigBee technology is its ability to perform mesh networks. While a typical ZigBee radio node may only have a range of 30 meters or less, the range can be extended by simply transmitting data from one node to another in a mesh network that may cover hundreds of meters, or in some cases even miles. A mesh is shown in Figure 9.14. Each circuit is a ZigBee transceiver node and each line is a wireless path. Mesh networks have many different paths for data and, therefore, are extremely reliable. If one path is blocked or disabled, data can usually find its way to the desired destination by way of another path. For example, in Figure 9.14, a direct path may be from A-D-F-I-K, but if the D-to-F link is blocked or the I node is disabled, the mesh would automatically reroute the data through another path such as A-D-G-J-K.

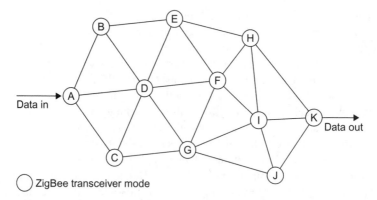

FIGURE 9.14   A mesh network can extend the overall range of a node and provide greater reliability through multiple possible paths. Most wireless mesh networks are ZigBee, but other wireless technologies may be used (e.g., Wi-Fi).

A forthcoming application of the 802.15.4 standard is the replacement of the current infrared (IR) TV remote controls. An RF version called RF4CE is gradually expected to replace IR remotes in all consumer electronic equipment providing longer range and no line of sight limitations. Two-way communications are expected to be implemented.

Ultra Wideband

Ultra wideband (UWB) technology is a very broadband, high-speed PAN designed for transmitting audio, video, and high-speed data. It uses OFDM in the frequency spectrum from 3.1 GHz to roughly 5 GHz. Higher frequencies up to 10.6 GHz are available but not widely used. Ultra wideband has a maximum data rate of 480 Mbps, but the actual rate depends on the range between nodes.

WiMAX

WiMAX is actually a wireless MAN. It was initially designed to offer a wireless alternative to Internet connections to homes, usually by way of cable TV systems or DSL lines of a telephone company. WiMAX is also standardized by the IEEE as 802.16. It comes in two variations, fixed and mobile. The fixed version is used for Internet connectivity in a fixed mode where the nodes in a star-like network are essentially in one position. WiMAX services are now available through an organization such as Clearwire in the United States with base stations with a range of several miles. Laptop computers containing WiMAX transceivers can connect to the network. Data rates will vary depending on the service offered by the company but can be from 1 Mbps to 10 Mbps or more. The technology uses OFDM with modulation schemes including BPSK, QPSK, 16-QAM, and 64-QAM. MIMO may or may not be used.

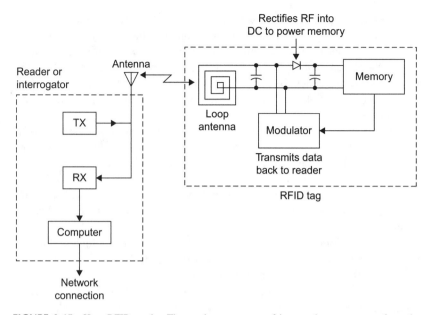

FIGURE 9.15   How RFID works. The tag has no power of its own but gets power from the transmitter in the reader.

A mobile version of WiMAX is also available. It is designed for connectivity where laptops and other mobile devices can connect to access points or base stations even while in motion.

Most WiMAX activity takes place in the United States in the 2.5-GHz frequency range and the 3.5-GHz range in Europe, Canada, and Asia. It is expected that WiMAX will be used in the new 700-MHz band for high-speed Internet access.

Radio Frequency Identification and Near-Field Communications

Radio frequency identification (RFID) is a very short-range technology that is intended to substitute for standard bar codes. Practically everything has a printed bar code on it today so that with optical scanners the device can be recognized by a particular part number or other information that identifies it. The bar code readers are normally attached to a computer and, in turn, a network.

RFID devices are small thin plastic tags containing a radio chip that can be read by a wireless reader. The RFID chip contains an EEPROM memory in which is stored the device's unique identification number and other related information. A key characteristic of the RFID tag is that it contains no power source such as a battery. Yet it is designed to transmit its digital code to some external reading device whenever it is interrogated. Figure 9.15 shows how it works.

The interrogation takes place when an external reader made up of a transmitter (TX) sends a radio frequency (RF) signal to the nearby tag. The radio frequency energy received by the tag's antenna is then rectified and filtered into a DC voltage that is used to power up the memory and the modulator circuit that sends data back to the reader. The reader also contains a receiver (RX) that picks up the small signal from the tag and sends it on to a computer where identification takes place.

RFID tags are generally used for what is known as asset tracking. RFID tags are attached to any relatively expensive item whose position is to be monitored, tracked, or otherwise controlled. It is good for keeping track of capital equipment, baggage, manufacturing tools, and other items. It is also used for personnel security on badges that employees wear and is good for animal tracking. One of the most widely used applications is automatic toll collection on highways.

RFID tags come in a variety of sizes, shapes, and technologies. Some of the older tags work at a frequency of 125 kHz or 13.56 MHz. The newer tags work at UHF frequencies in the 902- to 928-MHz range. All of these tags are passive devices with no power other than that received from the reader. Some longer-range RFID tags with small batteries have also been developed.

The unique code stored in the EEPROM is stored during the tag's manufacture. It is then clocked out at a relatively slow rate of speed around 70 kbps. The serial binary data is used in a form of amplitude modulation called backscatter modulation that produces minor variations in the reader's signal that can be converted back into a binary code.

Another similar RFID-like technology is known as near-field communications (NFC). It has an even shorter range of only about 1 foot. It operates on 13.56 MHz. It is mainly designed to be used in credit cards and in cell phones to produce automatic transactions such as credit card purchases or payment for transportation (e.g., train, subway, bus) or access to some facility.

BROADBAND TECHNOLOGY

Whenever you hear the term "broadband," it generally refers to a high-speed interconnection or Internet access. Most broadband systems provide data rates from roughly 1 Mbps or less to well over 10 Mbps, depending on the level of service purchased. Such broadband services are usually provided by cable TV companies or telephone carriers. Other smaller organizations also offer various other types of Internet connectivity.

Cable TV Connectivity

Most cable TV companies offer high-speed Internet connectivity services. What they do is allocate one or more of the 6-MHz wide channels on the cable to this service. (See Chapter 11 on TV.) By using QPSK, 16-QAM, or 64-QAM, a wide range of customers can be offered multiple levels of Internet connectivity service. The customer gets a cable TV modem that connects to the cable along with the TV set. This modem provides the connectivity to a computer via Ethernet. A Wi-Fi access point is often connected to the cable modem to provide wireless connections to the Internet for PCs and laptops.

Digital Subscriber Lines

Digital subscriber lines (DSLs) are high-speed Internet connections that use standard telephone lines. DSL uses the standard twisted-pair telephone lines that come into every home for normal telephone service. These lines, because they were developed only for voice signals, are restricted in bandwidth and data rate. However, special techniques have been developed to allow very high-speed data transmissions on them. DSLs, also referred to as *asymmetric digital subscriber lines* (ADSL), use a variation of OFDM called *discrete multitone* (DMT). What it does is divide the restricted bandwidth of the twisted-pair cable into multiple OFDM channels, each 4 kHz wide. Then the data to be transmitted is divided into parallel paths and modulated using some form of QAM. All of this is handled by a DSL modem connected to subscribers' computers and the home telephone lines.

The speed potential of an ADSL line depends on how far away the subscriber is from the central office. The greater the distance, the lower the data rate. For even the longest runs from 12,000 to 18,000 feet, data rates of up to about 2 Mbps are possible.

Newer versions of ADSL have also been developed to permit data rates of up to about 12 Mbps at a range of 8000 feet and 20 Mbps at a range of about 4000 feet. A newer version referred to as video DSL or VDSL extends the bandwidth further and uses higher-level versions of QAM to get data rates of up to 52 Mbps.

Wireless Broadband

Wireless broadband services are only beginning to become available. A relatively new WiMAX wireless network known as XOHM is offered by Clearwire, a spinoff of Sprint Nextel. It provides fixed wireless data services with data rates in the 1- to 3-Mbps range. WiMAX modems are connected to computers that access nearby base stations within several miles of one another.

The beauty of wireless broadband is that no cables or other hard connectivity is required. These are easy to set up, although at microwave frequencies, range is a problem as are multipath problems with buildings, trees, and other obstructions. However, with high enough antennas and sufficient power, wireless broadband has proved to be practical. It is growing rapidly, particularly in those areas where DSL and cable TV services are not currently available. This is particularly true in rural areas and in some small towns and communities.

Don't forget that wireless Internet connectivity can also be achieved by using the regular cell phone networks. You can purchase a data modem that plugs into your laptop USB connector on a PC. This modem connects you just as it would a cell phone to your cellular network. From there you can use the 3G data service to connect to the Internet, access email, or perform other networking operations.

ISM-BAND RADIOS

The industrial-scientific-medical (ISM) band has been mentioned before. This comprises a group of frequencies or bands set aside for unlicensed data transmission over short ranges. The main application is telemetry monitoring and remote control. The U.S. frequencies are 13.56 MHz, 315 MHz, 433 MHz, and 902 to 928 MHz, with 915 MHz the most common. In Europe, 868 MHz is an unlicensed frequency. Transmitters and receivers are tiny single ICs with power levels from about 1 mW to 1 watt. Modulation is usually ASK/OOK or FSK. Data rates are low from a few kbps to usually no more than 100 kbps.

Some typical applications are garage door openers, remote keyless entry for vehicles, remote thermometers, medical telemetry, remote-controlled toy cars and boats, and sensor reading in industrial applications. Range is no more than a hundred feet or so, but longer ranges to several miles can be achieved with higher antennas.

Project 9.1

Become More Familiar with Wireless Data Protocols
You may want to find more information on the most popular short-range wireless data methods. The following websites are recommended. And you may want to do a general Internet search with Bing, Google, or Yahoo!.
  Wi-Fi: www.wi-fi.org
  Bluetooth: www.bluetooth.com
  ZigBee: www.zigbee.org
  General sites: www.palowireless.com, www.howstuffworks.com
The Wikipedia entries are also quite good.

Chapter 10

Audio Electronics
Digital Voice and Music Dominate

In this Chapter:
- Sound and the electronics that deliver it.
- Digital audio.
- The compact disc (CD).
- Voice and music electronics.

INTRODUCTION

While the telephone, telegraph, and radio are generally considered the first real electronic applications, audio was certainly next. By "audio," I mean sound: voice and music, things we can hear. Audio signals have frequencies within the range of the human ear. Over the years many electronic devices have been developed to amplify, capture, and reproduce sounds of all types. Today, like everything else in electronics, audio electronics is predominantly digital, although many analog devices and systems are still in use. This chapter is a summary of audio devices you know and use every day from stereo systems to iPods.

THE NATURE OF SOUND

Sound is just pressure waves in the air that your ear can hear. Anything that produces a noise or other disturbance produces pressure waves that travel to your ears where the sound waves are converted into signals that your brain recognizes as sound. The frequency range of these air vibrations is 20 Hz to 20 kHz. Our ears do not respond to lower or higher frequencies even though they may be present. Human hearing also varies widely due to age and health status. Age brings on hearing loss, especially of the higher frequencies.

There are four key things needed in audio electronics:

- The ability to convert sound waves into electrical signals.
- The means to convert the electrical signals into sound.

doi: 10.1016/B978-1-85617-700-9.00010-2

- A way to amplify the signals.
- Media to store or record the signals.

This is what audio electronics is all about.

Microphones

The microphone is the transducer that converts sound waves into an analog signal. The sound waves impinge on the microphone and it causes vibrations that are translated into a signal by either capacitive element or an inductive coil. A technical discussion about microphones is way beyond the scope of this book. In any case, the resulting signal is usually very small, in the millivolt range or even microvolt range. Therefore, it must usually be amplified before it is useful. This is the function of a small-signal amplifier usually called a *preamplifier*. What you do with the signal after that is up to you. Typically it will be amplified further for use in a public address (PA) or other sound system. Or alternately, the voice may be recorded. Many voice signals will be used in a telephone, cordless or cellular.

Speakers

A speaker is a transducer that converts the audio signal into sound waves that you can hear. You have obviously seen a speaker, so little explanation is needed. Its main element is a paper or plastic cone that vibrates at the audio frequencies applied to it. Then it produces the sound pressure waves that our ears will hear.

Figure 10.1 shows how a speaker works. The cone is attached to a form around which a coil of wire is wound. This is the so-called *voice coil*. That voice coil is positioned inside a strong permanent magnet with north and

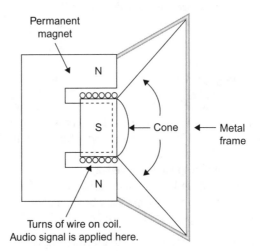

FIGURE 10.1   Cross-section of speaker showing major components and how it works.

south poles. Now when you apply the audio signal to the coil, current flows and causes the coil to become an electromagnet. The electromagnet generates alternating north and south poles as the voice or music signal occurs. These magnetic fields interact with the permanent magnet, causing attraction and repulsion of the voice coil, and the cone moves. The motion of the cone accurately reproduces the original sound waves.

Speakers come in a wide range of sizes and types. Cone speakers are usually called mid-range speakers in that they can reproduce most of the frequency range except for the very low and very high frequencies. Special speakers are used for those frequencies. A woofer or subwoofer is a larger speaker designed for frequencies of about 600 Hz and below. A tweeter is a special speaker optimized for the higher frequencies above 10 kHz. Special filters called *crossover networks* separate the signal into low, mid-, and high frequencies as shown in Figure 10.2.

The remainder of this chapter addresses the amplification and recording or storage of audio signals. Along the way, most analog audio signals are digitized. And many are also compressed to save storage space or reduce the binary data rate for transmission.

DIGITAL AUDIO

The sounds we hear and the sounds we generate are analog signals. A microphone picks up the voice and music and produces an analog signal to be amplified or stored. We have amplifiers and speakers that will accurately reproduce the

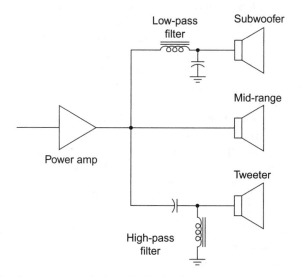

FIGURE 10.2   Cross-over networks keep signals of different frequencies separate for appropriate speakers.

sounds. But the real challenge with audio is capturing, storing, or recording the sounds for later reproduction. With today's digital techniques, we can store not only voice and music more accurately than before, but we can also transmit it more easily by radio, TV, or the Internet.

Past Recording Media

Edison invented the phonograph in the late 1800s. This was the first device ever conceived to record voice. A microphone converted the sound into vibrations that moved a sharp needle in direct accordance with the sound. The needle cut a varying groove in the wax coating on a rotating cylinder. Later, a needle placed in the groove moved the diaphragm in a horn speaker that translated the movement into sound. The quality was poor but it worked, and the concept was soon translated into what we know as a *phonograph record*. The record disc is rotated while a needle cuts a spiral groove into the plastic that accurately captures the sound. To reproduce the sound the disc is rotated at the same speed and a needle is placed in the groove to reproduce the sound in a speaker. A transducer converts the movement of the needle in the groove into an analog signal that is amplified.

Early records spun at a speed of 78 rpm and could only hold a few minutes of sound. Later a smaller 45-rpm record was invented to play minutes of music at a lower cost. Then a larger long-playing record spinning at only 33⅓ rpm was produced that could hold an hour or so of sound. Stereo or two channel sound was also effectively recorded into a V-shaped groove in the record.

Records were soon upstaged by magnetic tape. A plastic such as Mylar was coated with a powdered iron or ferrite magnetic material. The music or sound was then applied to a magnetic coil that put a magnetic pattern on the tape representing the sound. A coil passed near the tape would convert the magnetic pattern variations into a small voltage to recover the sound. Reel-to-reel tape was popular for a while, but magnetic tape became the recording media of choice for years when the Philips cassette tape was invented. Other formats like the ever-popular 8-track unit were also used.

All of these formats worked well but had their limitations. Record grooves wore out with usage causing the high-frequency response to be eroded. Scratches and dirt on the record also produced noise. As for tape, it had good frequency response but had its own noise that could not be reduced below a certain level. The dynamic range of both was poor.

Digital technology has solved all these problems. Now the most popular medium is the compact disc (CD), which records sound in full digital format. It has very low, practically undetectable noise, wide frequency response, and a very long life. The cost is also low.

Digital sound is also routinely stored in computer memories like flash EEPROMs and transmitted over the Internet. With digital techniques, sound can also be processed with DSP such as compression, filtering, and equalization.

Dynamic Range

Dynamic range refers to the difference between the highest and lowest volumes of sound that a system can handle. It is a measure of how wide an amplitude range a particular storage media can record or an audio system can accommodate. It is also a way to state the maximum sound pressures that our ears can withstand and the lowest levels that we can hear.

The upper level of amplitude is set by our ears and the level at which distortion occurs. The lowest levels of the dynamic range are determined by the noise in the system. Noise is the random voltage variation caused by thermal agitation of electronics in electronic components and random variations in the magnetism on a magnetic tape or disc. As the volume of a sound gets lower at some point it will be smaller than any noise present and not distinguishable from the noise background.

In general, the wider the dynamic range the better. It usually means lower noise and greater volume (loudness) to better match the range of human hearing.

Dynamic range can be stated in terms of power or voltage. It is usually a very wide range, so we resort to a unique mathematical method to express it. The term decibel (dB) refers to how dynamic range is presented. In math form you take the ratio of the maximum and minimum voltage, take the logarithm of that, and multiply by 20:

$$dB = 20 \log (V_{max}/V_{min})$$

For power,

$$dB = 10 \log (P_{max}/P_{min})$$

The dynamic range of a typical phonograph record is probably about 50 dB, and magnetic tape, about 55 dB. But a compact disc (CD) is approximately 90 dB, the best we have.

Dynamic range can be improved by lowering the noise in a system. This can be done with special signal-processing techniques. The most notable are methods produced by Dolby.

Digitizing Sound

Most digital sound and music is digitized with a sampling rate of 44.1 kHz or 48 kHz. Recall the Nyquist requirement that to capture and retain the frequency content accurately, the analog signal has to be sampled at least twice the highest frequency of the sound. Since most music and the human ear cut off at about 20 kHz, then any sampling rate greater than 2×20 kHz $= 40$ kHz will satisfy that requirement. The rate of 44.1 kHz was standardized for CDs and 48 kHz for professional recordings. At 44.1 kHz, a sample is taken every $1/44,100 = 22.676$ microseconds.

Figure 10.3 shows the sampling of an analog signal. The smooth curve is the analog music or voice. At each vertical interval a sample is taken. In this

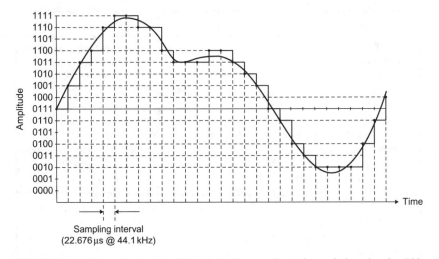

FIGURE 10.3   How audio signal is digitized. In this example, each sample is assigned a 4-bit code. The stepped approximation shown is what a DAC would deliver to the speakers.

example, a 4-bit ADC is used so there are $2^4 = 16$ possible levels. Note the 4-bit code associated with each sample amplitude. The binary samples are stored for later playback. When sent to a DAC, the samples will reproduce the audio in a stepped approximation shown overlaid with the analog signal. Some low-pass filtering of that signal will produce a sound which almost duplicates the original.

Digitizing audio usually results in each sample producing a 16-bit word. With 16 bits you can represent $2^{16} = 65,536$ levels. That produces a dynamic range from the highest to lowest amplitude levels of 65,536 to 1. On the decibel (dB) scale, that is a range of 96 dB—far greater than any dynamic range of previous analog recording methods. And noise is virtually nonexistent. It is so low no one can hear it. The overall result is a highly accurate way to capture sound and music. The 16-bit words can then be conveniently stored in a memory, or captured on a compact disc.

Digital Compression

Digital compression is a mathematical technique that greatly reduces the size of a digital word or bitstream so that it may be transmitted faster or stored in a smaller memory. Digitizing sound creates a huge number of bits. Assume stereo music that sampled at a rate of 44.1 kHz to create 16-bit words for each sample. One second of stereo music, then, produces $41,000 \times 16 \times 2 = 1,411,200$ bits. A 3-minute song is $60 \times 3 = 180$ seconds long. The result is $1,411,200 \times 180 = 254,016,000$ bits. Since there are 8 bits per byte, the result is 31,742,000 bytes or nearly 32 MB or megabytes. That is an enormous amount of memory for just one song. With a recording medium like the CD with a storage capacity of about 700 MB, that is okay. But for computers or portable music devices, it is impractical, not to mention expensive. And to transmit that over the Internet would take about 4 minutes at a 1 Mb/s rate. Pretty slow by today's standards.

The solution to this storage and transmission problem is to compress the bitstream into fewer bits. This is done by a variety of mathematical algorithms that greatly reduce the number of bits without materially affecting the quality of the sound. The process is called *digital compression*. The music is compressed before it is stored or transmitted. Then it has to be decompressed to hear the original sound.

The two most commonly used music compression algorithms are MP3 and AAC. MP3 is short for MPEG-1 Audio Layer 3, the algorithm developed by the Motion Picture Experts Group as part of a system that compressed video as well as audio. AAC means advanced audio coding. MP3 is by far the most widely used for storing music in MP3 music players and sending music over the Internet. AAC is used in the Apple iPod and iPhone and used on the iTunes site to send music. It is also part of later MPEG2 and MPEG4 video compression formats. Both methods significantly reduce the number of bits to roughly a tenth of their original size, greatly speeding up transmission and easing storage requirements. There are many more compression standards out there, but these are by far the most used and the ones you will most likely encounter.

To perform the compression process you actually need a special CPU or processor. It is typically a special DSP device programmed with the algorithm for either compressing or decompressing the audio.

There are also a number of compression methods used just for voice. Voice compression was created to produce signals for telephony. Most phone systems assume a maximum voice frequency of 4 kHz. The most common digitizing rate is twice that or 8 kHz. Eight-bit samples are typical. If you digitize voice creating a stream of samples in serial format, the signal would look like that in Figure 10.4. Each sample produces an 8-bit word where each bit is $125/8 = 15.625\,\mu s$ long. That translates to a serial data rate of $1/15.625\,\mu s = 64\,kbps$. This takes up too much bandwidth in a telephone system, so compression is used. The International Telecommunications Union (ITU), an international standards organization, has created a whole family of compression standards. These are designated as G.711, G.723, G.729, and others. These mathematical algorithms reduce the bit rate for transmission to about 8 kbps. You will see them used in VoIP (voice over Internet phone) digital phones, which are gradually replacing regular old-style analog phones.

There are many other forms of audio compression. Another common one is Dolby Digital or AC-3 that is used in digital movie theater presentations and some DVD players.

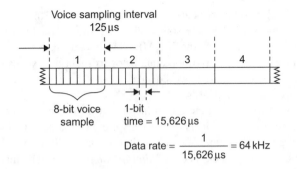

FIGURE 10.4   Voice signals up to 4 kHz for the telephone are sampled at an 8-kHz rate, producing an 8-bit sample, each 125 microseconds. Stringing the samples together in a serial data stream produces a digital signal at a 64-kHz rate.

How an MP3 Player Works

MP3 players such as the Microsoft Zune or Apple iPod are amazing electronic systems in a hand-held package. They all take the general form shown in Figure 10.5. At the heart of the player is a processor. It may be a general processor or a DSP. In some cases, two processors are used, one to control the player buttons, controls, and display, and other general housekeeping functions; a separate DSP handles the compression/decompression. A ROM holds the general control program for the device. A large flash EEPROM is used to store the music in compressed form. Some devices also provide for an external flash memory plug-in device to hold more music. A USB port is the main I/O interface to connect to a PC for the music downloads.

The codec in the device is the coder-decoder. This chip holds the ADC to digitize music from an external source such as a microphone (coder) and the DACs that decode or translate the decompressed audio back into the original analog music for amplification. Class-D switching power amplifiers are used for the headphones to reduce power consumption. One or more internal speakers may also be used. Other circuits include the LCD display and its drivers and LED back-lights if used. A power management chip is normally used to provide multiple DC voltages to the different circuits and to manage power to conserve it. This section also contains the battery charger.

The Compact Disc

The compact disc (CD) is by far the most popular digital audio medium in use today. It has been around since the 1980s and is still going strong. It has virtually totally replaced the Philips cassette magnetic tape cartridges and vinyl phonograph discs. About the only competition the CD has is from flash EEPROM devices such as those used to store MP3/AAC music for portable music players.

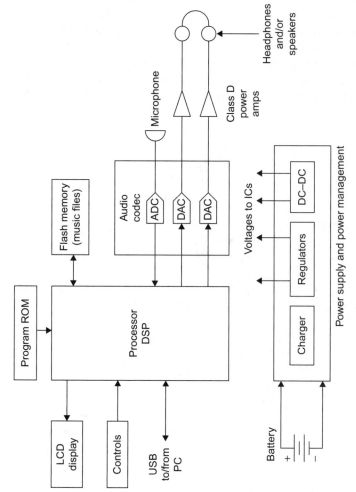

FIGURE 10.5 General block diagram of a digital music player like Zune MP3 or Apple iPod.

You have probably seen a CD. Let's add a little background so you will appreciate the CD. It is a 4.72-inch (120-mm) diameter disc made of plastic. It is about 0.05 inch (1.2 mm) thick and has a hole in the center for mounting on the motor that spins the disc. It is actually a sandwich of clear plastic and a thin metal layer for reflecting light.

The music is recorded on the disc in digital format. A laser actually burns *pits* into the plastic representing the 1's and 0's of the digital data (see Figure 10.6). These pits are extremely small, about 0.5 μm wide (a micron designated μm is 1-millionth of a meter). The flat areas between the pits are called *lands*. A binary 1 occurs at a transition from a pit to a land or vice versa while the land is a string of binary 0s. The binary data is recorded as a continuous spiral track about 1.6 μm wide, starting at the center and working outward. That is a very efficient way to store the data, but it means that the motor

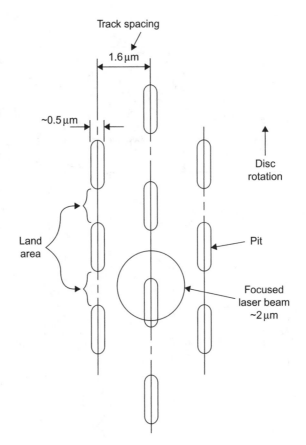

FIGURE 10.6   Digital music on a CD is encoded as pits and lands on the CD surface. These are burned in by a laser. A laser light shined on the pits and lands produces a reflection that a photodetector picks up to create the original digital data.

speed must vary to ensure a steady bit rate when the music is played back. The pickup mechanism will experience higher speeds at the center of the disc and lower speeds at the outer edges of the disc, so that a speed control mechanism keeps the recovered data rate constant.

The music on a CD is derived from two stereo channels of music with a frequency response from 20 Hz to 20 kHz. Digitization occurs at the standard 44.1-kHz rate. Each sample is 16 bits long. The 16-bit words from the left and right channels are alternated and formatted into a serial data stream that occurs at a rate of 44.1 kHz $\times$ 16 $\times$ 2 = 1.4112 MHz. The 16-bit words are then encoded in a special way. First, they undergo an error detection and correction encoding scheme using what is called *cross-interleaved Reed-Solomon code* (CIRC). This coding helps detect errors in reading the disc caused by dirt, scratches, or other distortion. The CIRC adds extra bits that are used to find the errors and fix them prior to playback.

Next, the serial data string is then processed using what is called eight-to-fourteen modulation (EFM). Each 8-bit piece of data is translated into a 14-bit word by a look-up table. EFM formats the data for the pit-and-land encoding scheme. Finally, the completed data is formatted into frames 588 bits long and occurring at a rate of 4.32 Mbps. That is the speed of the serial data coming from the pickup assembly in a CD player before processing.

To recover the audio from the CD, the disc is put into the player and a motor rotates it. Refer to Figure 10.7. A laser beam is shined on the disc as indicated in Figure 10.6. The reflections from the pits and lands produce an optical light pattern that is picked up by a photodetector that converts the light variations into the 4.32 Mbps bitstream. A motor control system varies the speed of the motor to keep the data rate steady. The data stream goes to a batch of processing circuits. First, the EFM is removed, and then a CIRC decoder identifies and repairs any errors and recovers the original 1.41-Mbps data stream. A demultiplexer separates the left- and right-channel 16-bit words and sends them to the DACs, where the original analog music is recovered and sent to the power amplifiers and speakers.

The CD can store lots of data. Its capacity is in the 650- to 700-MB range. That translates into a maximum of about 74 minutes of audio. And this is not compressed.

AV RECEIVER

An HDTV receiver is usually at the heart of most consumer electronics home entertainment systems. A secondary piece is the AV (audio–video) receiver, which provides the audio component of the entertainment. The systems usually include the CD player, several radio options, and all the audio power amplifiers that operate the multiple speakers. The AV receiver also accepts inputs from the TV set, DVD player, and other external devices to provide higher-quality sound. This section provides a look at this piece of equipment and how it works.

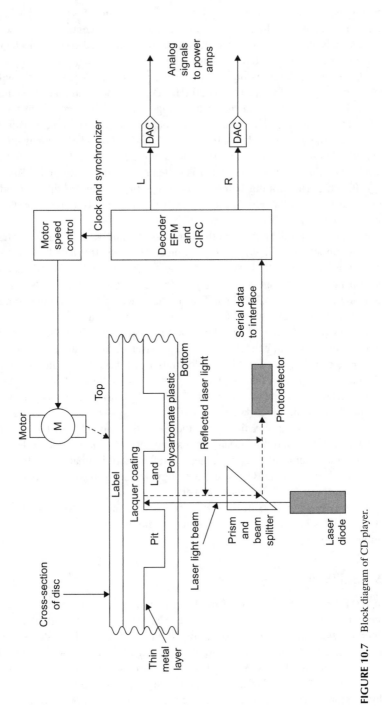

FIGURE 10.7   Block diagram of CD player.

Figure 10.8 shows a general block diagram of the AV receiver. Note the multiple power amplifiers on the right. These drive the speakers that are part of a surround sound system. Most of these are class AB linear amplifiers that deliver power levels from about 20 watts per channel to over 100 watts in the larger units. The frequency response is 20 Hz to 20 kHz or more with a distortion level of 0.5 to 0.9% or less.

The power amplifiers get their inputs from multiple sources. These sources include audio from the internal radio receivers and the CD player as well as many other external sources. These sources include the TV receiver, DVD player, VCR, and a cable or satellite box, as well as audio from older devices like tape players and phonographs. Many receivers also accept inputs from an iPod or other MP3 music player. A large switching matrix is used to select the

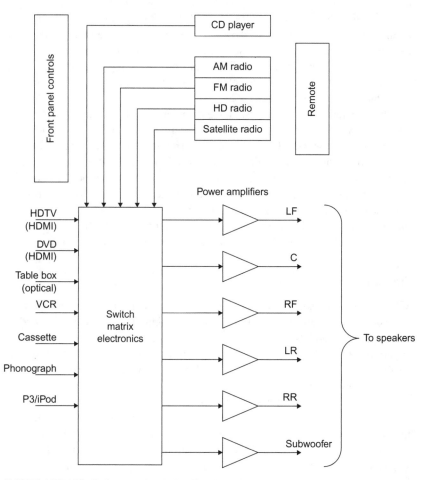

FIGURE 10.8   Block diagram of typical AV receiver, the heart of the audio reproduction part of a consumer home entertainment center.

desired input. That selection can be made from the front panel controls or via the remote control.

Radio Choices

All AV receivers contain the traditional AM and FM analog radios. These are still popular sources of music. But more and more the most recent receivers contain one or more digital radio sources such as HD radio or satellite radio.

HD radio is becoming more popular each year. This is a digital radio broadcast option available at most U.S. radio stations today. The AM and FM sources are digitized and broadcast on the same frequencies but with digital modulation techniques (OFDMs). The result is an improvement in the signal. Digital signals offer slightly better frequency response on both AM and FM. AM signals will sound more like FM and FM signals will be nearly CD quality. Furthermore, the digital signals are more noise- and fade-free than the standard AM and FM signals. While HD radio was first made available in automobiles, it is now widely available for home use. Separate HD radios can be purchased at most consumer electronics stores. And more HD radios are being built into AV receivers.

Satellite radio was also first made popular in automobiles. Yet separate home receivers are available. Some AV receivers now include satellite radio capability. Satellite radio has been available from two sources, Sirius Radio and XM Radio. Both are subscription services with hundreds of radio, music, news, sports, and other channels. The signals are digital, delivering near CD-quality sound. Both systems use the 2.3-GHz microwave band. The main difficulty when using satellite radio at home is that an outside antenna is needed. Some radios supply a small antenna for window mounting so that the antenna can "see" the satellites in the sky. Otherwise, poor or no reception may occur.

Surround Sound

Virtually all audio sound today is stereo. That is, the music comes from at least two speakers, one on the left and one on the right. The sound is recorded as two separate channels with different microphones to provide a stereophonic experience to the listener. A minimal system uses two speakers but some add a subwoofer for good bass reproduction. This is sometimes referred to as a 2.1 stereo system, that is, two speakers and one subwoofer.

New audio systems offer surround sound. These systems feature six, seven, or eight speakers. The idea is to totally envelop the listener with sound as it might be heard in a concert hall or theater. There are speakers in front and at the rear. The first surround-sound implementations were analog but today most surround sound is fully digital. The most common surround-sound method is Dolby Digital 5.1, which uses five speakers and a subwoofer as shown in Figure 10.9. These are left front (LF), right front (RF), center (C), left rear (LR), and right rear (RR). Then there is one subwoofer. The woofer can be positioned

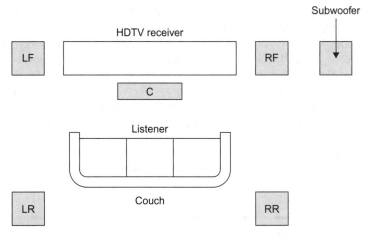

FIGURE 10.9   How 5.1 surround sound system is set up.

anywhere but is usually at the front. Each speaker is driven by its own amplifier in the AV receiver.

The big question that you may have is where do the six channels of sound come from? Not from the radio or a CD for sure, both of which are stereo only. In most cases, the sound signals will come from a DVD player. Right now it is about the only common source of 5.1 surround sound, although some HDTV programs support it. During recording of the original material, one microphone is used for each channel. The digitized audio for each of the channels is stored on the DVD.

Incidentally, there are more elaborate versions of surround sound. For example, 7.1 adds two more speakers on the left and right. As of now there are few, if any, 7.1 music sources available.

When you play the DVD, the digital signals from each channel are sent to a special chip that decodes them and otherwise performs necessary noise reduction, equalization, and other operations. The recovered signals are then sent to DACs for conversion to their original analog format for amplification.

SPECIAL SOUND APPLICATIONS

When you think of audio or sound, you think mainly of voice or music reproduction. Yet, electronic sound is used in lots of additional places. Some interesting examples are in musical instruments, sonar, ultrasound, and hearing aids.

Musical Instruments

Musical instruments produce their own sounds, of course. Each has its own unique signature made up of the basic notes or tones to which are added the mix of harmonics or overtones unique to the physical structure of the instrument. Some instruments like horns and drums rarely need amplification. Other instruments like guitars and violins commonly need some amplification, especially in large venues such as on stage addressing hundreds or thousands of people. Special microphone-like devices called pickups are used with string instruments. The pickup is usually one that converts the string vibrations into a small signal that is then amplified.

One of the most popular electronic instruments is the *synthesizer*. It has a keyboard like a piano. The keys are equivalent to those on a piano. Each key then triggers the electronics to generate a tone of the desired frequency. That tone is especially modified by DSP circuits to sound like that note played on a piano, an organ, or any other musical instrument. The DSP synthesizes or constructs the tones digitally and sends them to a DAC that converts them into the analog sound to be amplified.

Are Vacuum-Tube Amplifiers the Best Audio Amplifiers?

Vacuum tubes? I'm kidding, right? Actually, no. There are many audio experts, especially guitar players and some recording studio engineers, who still believe that vacuum-tube amplifiers are superior to solid-state amplifiers. And as it turns out, the largest market for general vacuum tubes today is guitar, high-end stereo, and recording studio amplifiers.

Contrary to popular belief, vacuum tubes did not really go away when transistors and integrated circuits came along. They are still widely used in China, Russia, and a few other countries. And that is where most tubes are made these days. Vacuum tubes are still widely used. Cathode ray tubes (CRTs) or picture tubes are still at the heart of many TV sets and computer video monitors. Magnetrons are vacuum tubes that are in every microwave oven. And microwave vacuum tubes, such as traveling-wave tubes (TWTs) and klystrons, are still used in high-power microwave transmitters. And vacuum-tube audio amplifiers.

Why are vacuum-tube amplifiers so popular still? It is the unique sound they produce. It is a sound that musicians appreciate and want in their sound mix. And the sound is difficult and expensive to reproduce in a solid-state amplifier. The sound is smoother and less harsh. The sound is "warmer," and the compression that such amplifiers produce when overloaded is exactly what makes rock and country music sound like it does. It looks like vacuum tubes will be around much longer, but prices are sky high and some tube models are hard to find.

Noise-Canceling Earphones

You have probably heard of these or maybe seen or used them. They make clever use of analog signal processing to greatly minimize surrounding noise when you are trying to listen to music with headphones.

First, most of these headphones try to isolate the actual headphone from the surroundings with a tight-fitting ear bud or soft foam pads around the earphone that limit the amount of external noise getting to your eardrum. This is called *passive noise control*.

Second, each earphone contains a built-in microphone that picks up the surrounding noise. It amplifies it then inverts it in phase. The original noise from the microphone is then added to the inverted noise. Being the same but exactly out of phase with one another, the two noise signals cancel one another. The cancellation is not perfect, but the noise reduction is significant. It can greatly minimize noise on a plane or in a car or other types of noisy environments.

Sonar

Sonar means sound navigation and ranging. It is basically the underwater equivalent of radar. It is used on ships and submarines for making depth measurements, detecting underwater objects, and for warfare. The two types of sonar are passive and active.

Passive sonar is simply the idea of putting underwater microphones called *hydrophones* on a long wire or arrays of such microphones in the water to listen for any sounds that occur. Amplifiers boost the signal levels and various filters help sift through all the complex signals that can be heard. In military applications, the sounds are digitized and various DSP programs help filter and identify specific sounds. Many ships create unique sounds called signatures that can be digitized and stored for comparison to any sounds picked up for positive identification. The sounds include fish and other sea life, boat and ship engines, motion through the water by a ship or boat, or any noise that occurs inside a submarine. Sound travels very easily through water so is easy to pick up. With practice and experience, a sonar operator can identify almost everything occurring over a wide range around the microphone array.

The other form of sonar is *active*. This is like radar where a high-frequency pulse of sound is applied to a transducer to convert the pulse into sound waves for transmission through the water. The frequency of the pulse is usually ultrasonic, that is, above most human hearing. Frequencies from about 15 kHz to 1 MHz are used, depending on the application. The pulse of sound travels out from the transducer and is then reflected off distant objects. The reflections go back to the source where they are picked up by a microphone. Then, knowing the speed of sound transmission in water, the distance to the object can be computed. Sonar depth sounders point the transducer downward to get a measure of the distance to the bottom. Sonars can easily pick up reflections from ships miles away for navigation safety purposes or for military actions.

Hearing Aids

Electronic hearing aids have been around a long time. The older analog types were large and cumbersome but worked fine. Integrated circuits made them much smaller and more acceptable to the hearing impaired. Today, the newer hearing aids use digital technology. The microphone picks up the sound and amplifies it then digitizes the sound in an ADC. A DAC can then translate the sound back into audio for amplification and application to the earphone. Most digital hearing aids now incorporate DSP filters. These filters may be customized to the frequency response of the ear, correcting for different frequencies. Mostly the high-frequency response is lost due to aging. Using an external computer, each hearing aid can be adjusted from volume and frequency response to fit the exact deficiency.

Project 10.1

Experience Digital Radio

Digital radio in the United States is available from satellites and via the existing AM and FM analog stations through HD radio. Both offer commercial-free music and other content in multiple channels. HD radio is free but satellite radio is a pay-for-listen service.

In the United States, two satellite radio services were established in the early 2000s, XM Radio and Sirius Radio. Recently the two services merged to form Sirius XM Satellite Radio. Both offer subscription services to music, news, sports, and other information. Both services operate in the 2.3-GHz band, and get their signals from satellites. Special receivers and antennas are required.

Satellite radio is a digital service and both systems used compressed audio for signal transmission. HD radio is simultaneously broadcast with existing AM and FM analog signals. A special receiver is needed to receive it. The digital techniques greatly improve sound quality.

A good project if you are a music lover is to try one or both of these services. You can get an HD or satellite radio for your vehicle or you can buy a home unit. Some high-end AV receivers have one or the other in addition to the traditional AM and FM radios. You can purchase home desktop units from Best Buy and other electronics dealers for a reasonable price.

HD radio is free and you will find that some stations transmit additional digital channels not available in analog form. These give you more music genres. No special antenna is needed.

To learn more about HD radio, do a Bing, Google, or Yahoo! search. And go directly to the company that invented HD radio, iBiquity (www.ibiquity.com), for more details.

You can also buy a satellite radio from several local sources. But you will need to subscribe. Go to www.sirius.com and www.xmradio.com.

Project 10.2

Test Your Hearing Response

To check the frequency response of your ears, go to www.audiocheck.net/audiotests_frequencycheckhigh.php. There are a few online audio tests for your hearing. This website has a whole batch of other audio response tests that you may be interested in.

Chapter 11

Video Technology

TV and Video Is All Digital Now

> **In this Chapter:**
> - How scanning works to create video.
> - Basic color TV principles.
> - U.S. and other standards for digital and high-definition TV.
> - How a digital TV set works.
> - TV screen technologies.
> - Cable TV.
> - Satellite TV.
> - Cell phone TV.
> - DVD players.

INTRODUCTION

Nothing in electronics has affected us more than the development of television. Okay, I haven't forgotten the PC and the Internet, which have also seriously impacted our lives. But TV came first and has been working its magic on us for far longer. The impact of television or video on our lives is significant in that we spend more than half our waking hours in front of a TV screen, either watching TV or working with a video monitor on a computer. The social and political impact has been enormous. And, of course, you also interact with the liquid crystal display (LCD) screens on cell phones, laptops, and personal navigation devices. For this reason it helps to understand a little bit about how video works. Video is perhaps one of the most complex segments of electronics, but this chapter makes those fundamentals easy to understand.

VIDEO FUNDAMENTALS FOR THE IMPATIENT

The main feature of any video device is the screen. The first video or TV screens were cathode ray tubes (CRT) and those are still used in older TV sets and computer monitors. The most commonly used TV and video screen today is the LCD. LCD screens are used in the newer TV sets and in most personal computers.

doi: 10.1016/B978-1-85617-700-9.00011-4

They are also used in laptops, netbooks, cell phones, personal navigation devices, and anything else that has a screen on it. TV sets also use plasma, digital light-processing (DLP) projection technology, and other methods to display very large pictures. There is more detail on TV screens later in this chapter, but for now, the most important part of understanding video is to learn the methodology used to create and present video regardless of the screen type.

A video system is a method for converting a picture or scene into an electronic signal that can be transmitted by radio or cable or stored electronically. Some type of device is needed to take in the light and color information from a picture or scene and convert it into an electrical signal. Once in electrical form, the video can be modulated onto a radio carrier for transmission or electronically stored. The process of converting the scene or picture into an electrical signal is known as scanning.

Scanning Fundamentals

A video signal is usually generated by a video camera. The camera takes the scene and scans it electronically, dividing it into hundreds or even thousands of very thin horizontal lines (see Figure 11.1).

The horizontal scan lines comprise a sequence of light and color variations. The camera produces an analog signal whose frequency and amplitude vary with the brightness and color changes along the scan line. The whole idea in video is to transmit the video signal one scan line at a time at very high speeds. If you are not transmitting the signal by radio or cable, you can store it, such as recording it on magnetic tape. In any case, the received signals are then converted back into scan lines and applied to the screen by scanning the lines across the receiving screen very fast from left to right and top to bottom.

The secret to producing a high-quality video picture is to use enough scan lines with enough detail in each so that the color, brightness, and fine detail in the picture are retained. Note in Figure 11.1 that the analog signal for one scanned line is such that the higher-amplitude voltages are closer to black while the lower-amplitude signals are closer to the white level. This is what a scan line from a monochrome or black and white (B&W) video system produces. It is called the *brightness* or *Y signal*. Color will be covered later.

Aspect Ratio

When talking about TV screens, the key factors are size and aspect ratio. The *size* is usually the diagonal measure of the screen in inches. The *aspect ratio* is the ratio of the picture width to the picture height. Standard TV sets and video monitors use an aspect ratio of 4:3 as shown in Figure 11.2. Most of the newer TV screens and video monitors have an aspect ratio of 16:9. This wider format was chosen because it is a better fit with the wide-screen movie formats used by Hollywood.

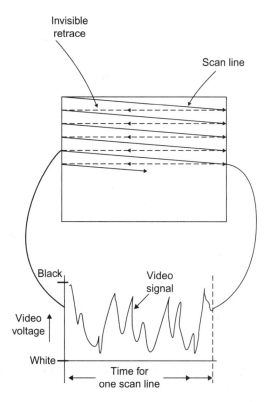

FIGURE 11.1   How scanning converts a scene of light and color into an electronic signal.

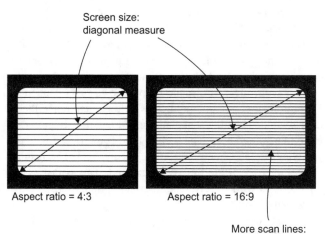

FIGURE 11.2   TV screen size and aspect ratio.

Persistence of Vision

When you think about it, it just doesn't seem possible that a complete picture or scene can be reproduced in electronic format by scanning. The scanning process just breaks the picture or scene down into hundreds or perhaps thousands of scan-line signals, and the only way that a picture can be created is for those scan lines to occur at a very high rate of speed. Luckily, this technique works because we all have a visual characteristic known as persistence of vision. Persistence is that characteristic of the eye which prevents it from following very high-speed changes in light variations. In other words, if the scan lines are traced across the screen at a very high rate of speed, your eye simply cannot follow them one at a time but instead puts the whole thing together and sees a single steady picture. If you scan at too low a rate, you will actually begin to see the scan lines and your eye will interpret this as a kind of flicker.

The concept of persistence of vision has been known for decades. Motion picture films take advantage of this characteristic of our eyes as well. Remember that movies are just a high-speed sequence of individual still pictures recorded on film. If you can get the frame rate up to a level of approximately 24 frames per second or faster, your eye will translate the changes from one still frame to the next as motion.

Analog TV

As you probably know, the Federal Communications Commission eliminated analog TV broadcasts in the United States as of June 12, 2009. But while analog TV has been replaced by digital TV, that doesn't mean that it isn't still around. You will still find analog TV in many other countries of the world, and it is also still widely used in closed-circuit television. In any case, it is such a widely used standard that understanding it is useful.

Back in the late 1940s and early 1950s, the United States created what is called the National Television Standards Committee (NTSC) TV standard. It basically specifies that a picture is made up of 525 horizontal scan lines, known as a video *frame*. As it turns out, one frame is actually made up of two fields of 262.5 scan lines. The scene or picture is first scanned 262.5 times from top to bottom, creating one field. This field is then transmitted at a rate of 60 fields per second. Then the scene is scanned again with another 262.5 lines and these are transmitted next. The second field is interlaced with the first field to form the complete 525 line picture. See Figure 11.3. The result is that you will see 30 complete 525-line frames per second. This is fast enough for persistence of vision to work nicely. The interlacing helps significantly in reducing any perceived flicker.

One thing that you may be wondering is how do you keep the TV receiver screen in step with the video camera or other video source? This is solved by creating synchronizing pulses that are transmitted before and after each scan line (see Figure 11.3). After one field is scanned, a larger group of synchronizing pulses occurs between fields to cause the receiver to start the scanning back at

Other World TV Systems

The U.S. NTSC analog system is used in Canada, Mexico, and Japan. But other analog systems are used throughout the world. In Europe, specifically the UK, Germany, Italy, and Spain, the PAL (phase alternation line) standard is used. It uses 625 scan lines with a frame rate of 25 frames per second. In France and in some other countries, a system called SECAM for Sequenteil Couleur a Memorie (or sequential chrominance and memory) is used. It too uses 625 scan lines and a 25-frame-per-second rate. PAL and SECAM systems have a better vertical resolution for picture detail because of the greater number of scan lines, giving a perceptibly improved picture over the NTSC system. Most of the other countries of the world have adopted one of these three systems.

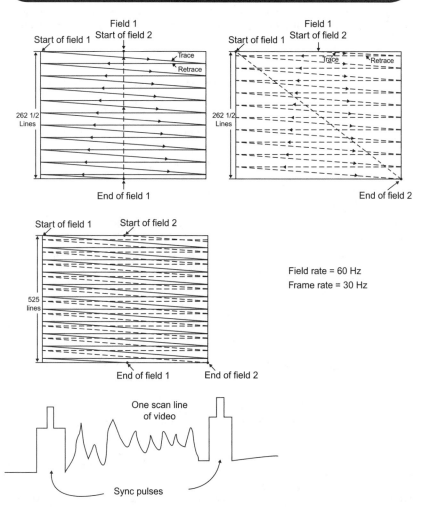

Field rate = 60 Hz
Frame rate = 30 Hz

FIGURE 11.3 Interlaced scanning helps reduce flicker. Sync pulses help keep the receiver and transmitter in step with one another.

the top again. The synchronizing pulses are transmitted or stored along with the analog video signal for each line. These synchronizing pulses are then used to trigger the receiver's circuitry to make sure that the scene is sequenced properly on the screen.

Digital TV

Most TV and video now is transmitted and stored in digital or binary format. Scanning is still the process used to convert the scene into an electrical signal, in this case a stream of binary numbers representing the brightness and color detail. Each scan line is still essentially an analog signal but that is converted into a sequence of binary numbers by an analog-to-digital converter (ADC). The basic process is illustrated in Figure 11.4. The ADC samples are measures of the analog signal at a constant rate, as shown by the dots along the curve

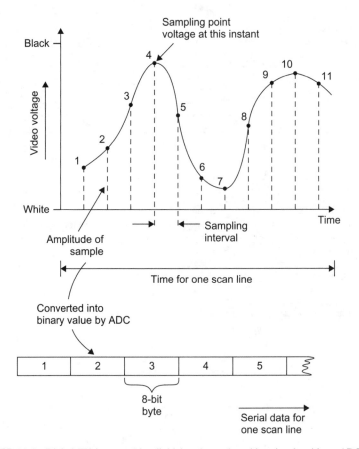

FIGURE 11.4 Digital TV is created by digitizing the analog video signals with an ADC, and then converting it into a serial bitstream of video data words.

in Figure 11.4. Remember that in order to retain the information in an analog signal when it is converted into digital, the signal must be sampled at a rate at least twice as fast as the highest-frequency signal contained in the analog signal. Since video signals contain extremely high frequencies, the sampling rate is very fast. Most video contains scene details with frequencies beyond 4 MHz. That means a sampling rate of at least 8 MHz to capture the detail. In any case, each sample results in an 8-bit binary number. These bytes, one representing each of the sequential samples, are sent one after another in a serial binary data stream as shown. The video data can then be transmitted or stored.

Each of the binary numbers produced by sampling produces what we call a *pixel* or picture element. This particular binary number will ultimately translate to a specific analog voltage value and produce a specific brightness level along the scan line.

One way to visualize a digital video display is to think of it as an array, or horizontal scan lines, each made up of a sequence of pixels (see Figure 11.5). Each pixel is a dot of light or, in this case, a square. That pixel is derived from a specific binary number developed during the scanning and sampling process. Each of those pixels represents a specific brightness level from white to black. A digital-to-analog converter (DAC) converts the binary brightness value back into a specific shade of gray.

Digital TV (DTV) uses the interlaced scanning principles described earlier to help minimize flicker. But it also uses a faster progressive scan. This is where each scan line making up the screen is scanned sequentially one after the other from top to bottom. To minimize flicker, *progressive scanning* must be done at a higher rate of speed than interlace scanning. Progressive scanning

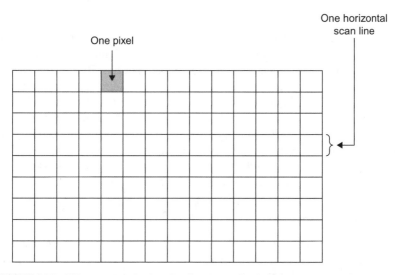

FIGURE 11.5   TV screens can be thought of as thousands of rows and columns of pixels, each represented by data.

is used simply because it creates a sequence of digital numbers that is more conveniently stored and processed one after the other than in a two-step interlaced process.

DTV uses a variety of different formats. In the basic format, it produces 480 scan lines with 640 pixels per line. This is roughly the equivalent of what you would see with an analog TV signal. This is what we call the DTV standard-resolution format. Both interlaced and progressive scanning formats can be used with rates of 24, 30, and 60 frames per second.

Picture Resolution

One thing that you should understand about video is the concept of resolution. This refers to how much fine detail you can see on the screen. In analog TV, brightness and color detail results in very-high-frequency analog signals. The older original analog TV was limited to frequencies of approximately 4.2 MHz. Modern HDTV uses much higher-frequency responses up to about 10 MHz. To see that picture detail, you must transmit the analog signal over a channel that has sufficient bandwidth. Most radio channels and cable systems have bandwidth limitations, and essentially act as a low-pass filter that will remove some of the higher frequencies. While you will still see a picture, it will lack some of the fine detail that your eye may or may not notice.

Resolution as it relates to DTV is a combination of how many scan lines plus how many pixels per line are used. Obviously, the greater the number of scan lines and the greater the number of pixels, the more detail you are going to be able to transmit and ultimately see.

A higher-resolution format to produce what we call high-definition (HD) TV uses 720 scan lines with 1280 pixels per line. The ultimate high-definition format uses 1080 horizontal lines with 1920 pixels per line. Both of those high-definition formats can use either interlace or progressive scanning at rates of 30, 60, 120, and 240 frames per second. Just remember that the frame rate simply refers to the number of times per second that each frame is presented on the screen. The earlier slow frame rates of 30 and 60 frames per second often produces a kind of picture smearing when the scenes are changing at a high rate of speed. This is noticeable in action movies or in sports events. The newer 120- and 240-frames-per-second scan rates essentially eliminate this problem.

Color Video Principles

Thus far, the signals we have described are what we call monochrome or black-and-white video. In *monochrome video*, all we capture and transmit and reproduce is the brightness variations along a scan line that range from black to white and an infinite number of shades of gray in between. But, of course, today most video is color. Capturing, transmitting, processing, and displaying color is a far more complex process than monochrome.

One physics principle that makes color video possible is the concept that any color of light can be reproduced by simply mixing the right combination of red, green, and blue light. Take a look at Figure 11.6. Mixing red and green creates yellow. Mixing red and blue produces violet (magenta), while mixing green and blue produces the color cyan. In any case, as it turns out virtually any color can be created using just the right mix and intensities of red, green, and blue. Strangely enough, there is one unique combination of the three colors that creates white. Black is produced, of course, by just turning off all the light sources. The color TV problem is how do you capture the individual colors and then how do you reproduce them?

Most video cameras use a light-sensitive imaging chip called a charge-coupled device (CCD). An optical lens focuses the scene on the chip. The chip itself is made up of millions of tiny light-sensitive elements that are charged like a capacitor with a voltage depending on the brightness level reaching them. The analog video signal for a scan line is then created by sampling those light-sensitive elements one at a time horizontally. The result is an analog signal whose amplitude is proportional to the light level. The result is a black-and-white brightness or Y signal.

Figure 11.7 shows how the color signals are produced. The scene is actually passed through a lens and a beam splitter that divides the scene into three identical segments. Each is passed through individual red, green, and blue filters. Three light-sensitive imaging chips (CCDs) then produce three separate analog signals for the red, green, and blue elements in a scene. The red, green, and blue analog signals are then individually amplified and sampled by an analog-to-digital converter (ADC) to translate them into a binary sequence. The three binary streams are then transmitted serially as a sequence of red, green, and blue for each picture element on the imaging chip. Each of these RGB colors will then later form a pixel on the viewing screen.

Video Compression

Generating a color video signal produces a serial bitstream that varies at a rate of hundreds of millions of bits per second. Consider these calculations:

- For each pixel, there are three color signals of 8 bits each making a total of $8 \times 3 = 24$ bits per pixel.
- For a standard DTV screen, there are 480 lines of 640 pixels for a total of 307,200 pixels per frame or screen.
- For one full screen or frame, there are $307,200 \times 24 = 7,372,800$ bits.
- If we want to display 60 frames per second, then the transmission rate is $7,372,800 \times 60 = 442,368,000$ bits per second, or 442.368 Mbps. That is extremely high frequency.

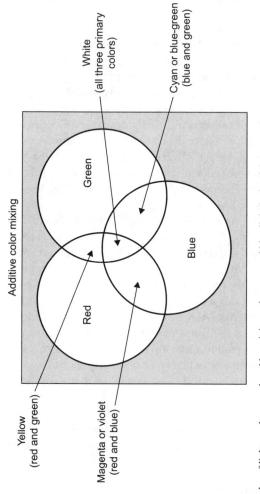

FIGURE 11.6   Any color of light can be reproduced by mixing red, green, and blue light in the right proportion.

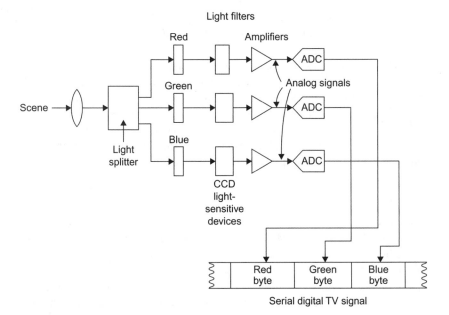

FIGURE 11.7 Red, green, and blue digital color signals are generated by filtering the scene into its red, green, and blue components, and then digitizing the resulting analog signals.

- To store one frame of video of this format, you would need a memory with 55,296,000 bytes or 55.296 MB. And that is just one frame.
- To store 1 minute of video you would need a memory of 3,317,760,000 bytes or just over 3.3 gigabytes (GB). And that is just 1 minute. Multiply by 60 to get the amount of memory for an hour.

Imagine the speeds and memory requirements for 1080p or a 120 frames-per-second format. And we have not even factored in the stereo audio digital signals that go along with this. Anyway, you are probably getting the picture here. First, data rates of 442.368 Mbps are not impossible, but they are impractical as they require too much bandwidth over the air and on a cable. And they are expensive. Furthermore, while memory devices like CDs and DVDs are available to store huge amounts of data, they are still not capable of those figures. Therefore, when it comes to transmitting and storing digital video information, some technique must be used to reduce the speed of that digital signal and the amount of data that it produces.

This problem is handled by a digital technique known as *compression*. Digital compression is essentially a mathematical algorithm that takes the individual color pixel binary numbers and processes them in such a way as to reduce the total number of bits representing the color information. The whole compression process is way beyond the scope of this book, but suffice it to say that it is a technique that works well and produces a bitstream at a much lower rate. And, the compressed video will take up less storage in a computer memory chip.

The digital compression technique used in digital TV is known as MPEG-2. MPEG refers to the Motion Picture Experts Group, an organization that develops video compression and other video standards.

What you have to think about when you consider the transmission of digital video from one place to another is that it is typically accompanied by audio. The audio is also in serial digital format. Those digital words will be transmitted along with the video digital words to create a complete TV signal.

TV Transmission

The compressed audio and video of a TV signal are assembled into packets or frames that look like that in Figure 11.8. The 4 bytes up front provide synchronizing bits for the receiver. The ID header tells the packet number, sequence, and video format used. The remaining bits are the serial compressed video and audio.

Figure 11.9 shows a simplified TV transmitter. The serial digital video is put through the MPEG-2 compression process. This is followed by additional

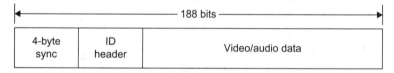

FIGURE 11.8   HDTV packet for ATSC U.S. digital color TV system.

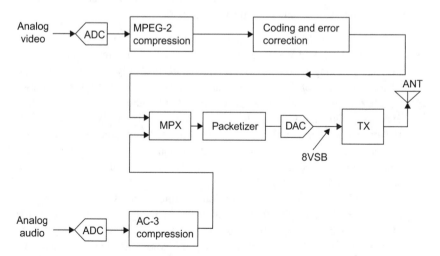

FIGURE 11.9   General block diagram of TV transmitter.

bit manipulation and coding for error correction and improved signal recovery by the receiver. This serial data is then multiplexed with the serial compressed audio data. Then the packets are formed. This is what will modulate the transmitter.

The modulation begins by translating the fast serial bitstream into an eight-level signal with a DAC. Each 3-bit segment of the serial data is assigned a specific voltage level as shown in Figure 11.10. Since each level represents 3 bits, you can transmit more bits per second in the same time period. It is this eight-level signal that modulates the transmitter carrier. The modulation is AM, where all of the upper sidebands are transmitted but only part of the lower sidebands. This saves spectrum. The modulation is referred to as 8VSB for *vestigial sideband*, where vestigial refers to part of the lower sideband. The signal is then up-converted to the final frequency by a mixer. A high-power amplifier in the transmitter sends the signal via coax cable to the antenna. Typical TV station range is about 70 miles maximum, and that is determined by tower height and terrain.

In the United States, each TV channel is 6 MHz wide. There are 50 channels starting at channel 2 from 54 to 60 MHz. Channels 3 through 6 range from 60 to 88 MHz. Channels 7 through 13 run from 174 to 216 MHz. Channels 14 through 51 run from 470 MHz and ending with channel 51 from 692 to 698 MHz.

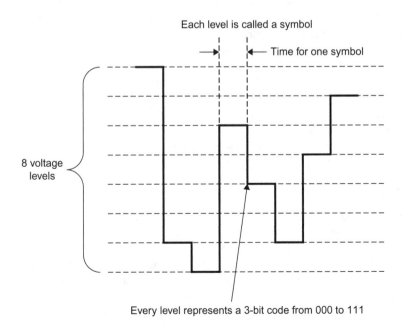

FIGURE 11.10 U.S. ATSC system translates the digital bitstream into an eight-level signal to transmit multiple bits per level, making higher data rates in a given bandwidth possible.

TV Receiver

Figure 11.11 shows a general block diagram of a modern digital TV receiver. The signal from the antenna or cable TV box is sent to the tuner where the signal is amplified and sent to a mixer along with a local oscillator signal to down-convert the signal to a lower frequency called the *intermediate frequency*. This is commonly 44 MHz. The local oscillator is a frequency synthesizer that selects the desired channel, usually with a remote control. The IF signal is filtered to select only the desired 6-MHz channel.

The IF is then sent to an ADC for digitizing. The 8VSB signal is then recovered in the demodulator. It is then translated into the original bitstream, and any coding or error correction is removed. The signal is then demultiplexed into the video and audio streams. The video goes to an MPEG-2 decompression circuit for recovery of the original video. The individual RGB components are sent to DACs to recover the picture detail. The RGB signals are then amplified and sent to the screen for display. The audio is also decompressed and sent to the DACs and amplifiers and speakers.

The whole TV set looks something like Figure 11.12. Note all the possible inputs. The set gets its RF inputs from the antenna or cable box. The cable box may also have a digital signal output, meaning the cable box does all the decoding, demultiplexing, and decompression. The output comprises a serial data stream of the video and audio. A special high-speed digital interface called the *high-definition media interface* (HDMI) is used. It uses a special cable no more than 15 feet long and with a special 19-pin connector. The HDMI carries video and audio from the TV to cable box or DVD player or vice versa.

Most TV sets also accept analog video inputs from other sources like VCRs, video cameras, and digital cameras. A special fiber optic cable input called TOSLINK serves as the interface for TV sets, cable boxes, and DVD players.

Finally, some TV sets include standard computer interfaces like USB ports or the IEEE 1394 interface, a fast digital serial interface for video and audio developed by Apple.

TV set outputs, of course, are sent to the screen and the speakers. A remote control sends signals to an embedded controller that tells the set what to do.

TV Screen Technology

As indicated earlier, there are several ways that the video signal is translated back into a viewable picture. The cathode ray tube is still present in older sets, but LCD screens are by far the most popular and most widely used today. Plasma screens are still available but slowly fading away, and DLP projection screens are popular because of their low cost. There are other emerging technologies that we may see in the future. In any case, what we need is a light source to reproduce the varying intensities of red, green, and blue light.

Color video screens are made up of a very large number of tiny red, green, and blue light sources. These are used to make up the pixels or dots of light

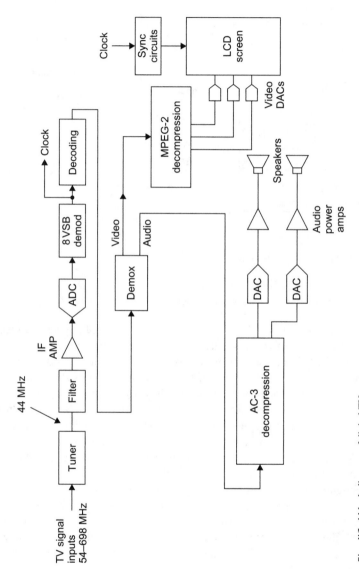

FIGURE 11.11   Simplified block diagram of digital TV set.

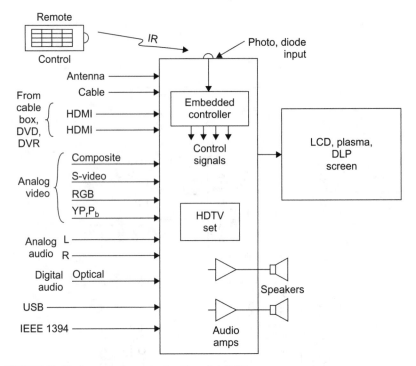

FIGURE 11.12   Inputs and outputs of modern digital TV set.

that re-create the picture. These pixels are very small, so that your eye cannot actually see them even when you are relatively close to the screen.

Figure 11.13 shows some of the screen pixel formats in use today. Cathode ray tubes were made up of a group of three color phosphor as shown in Figure 11.13A. These color triads were grouped tightly together to form the complete screen. When an electron beam strikes each of the red (R), blue (B), and green (G) pixels, they cause the phosphor dots to glow. The eye automatically mixes the individual red, blue, and green colors to reproduce the desired color.

LCD and plasma screens use a pixel format somewhat like that shown in Figure 11.13B. These are usually rectangular red, green, and blue areas that are backlighted with the appropriate degree of intensity.

One pixel is not necessarily equal to the three color group shown in Figure 11.13A. Instead, 1 pixel as displayed on the screen might actually be made up of a collection of these individual groups. The actual pixel size may be something like Figure 11.13B, made up of multiple groupings of red, blue, and green elements as shown.

The actual format displayed on the screen is determined by the TV display standards. As indicated earlier, a standard DTV screen uses 480 horizontal lines with 640 pixels per line. That creates a screen with a total of 307,200 pixels. The 720 × 1280 screen puts 1280 pixels on 720 lines for a total of

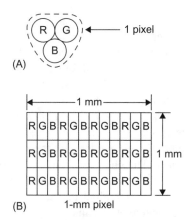

FIGURE 11.13   How color pixels are formed on a TV screen. (A) Pixel format for most CRTs. (B) Pixel format typical of LCD, plasma, or LED screens.

921,600 pixels. The big daddy of all screens uses 1080 lines with 1920 pixels per line for a total of 2,073,600 pixels. Just remember that the pixels are turned on one at a time from left to right and top to bottom on the screen at a very high rate of speed.

Summary of Screen Technologies

The actual technical details of each of the different types of screens are way beyond the scope of this book, but a summary might be useful. A quickie review of each of these screen types follows.

CRT—A cathode ray tube contains three electron-beam guns, each one driven by the analog color signals from the DACs. The intensity of the electron beam depends on the intensity of the light and color signal being fed to it. These electron beams are aligned so that they strike the red, green, and blue phosphor dots covering the screen. Each dot emits light with an intensity proportional to the strength of the electron beam striking it. The electron beams are scanned simultaneously across the screen at very high speed, stimulating the pixels in the right proportion to reproduce the desired color.

LCD—Liquid crystals are, as their name implies, a liquid made up of molecules whose physical orientation can be changed by the application of an electric signal. Different voltages change the polarization of the light through the liquid. By applying one level of voltage to the liquid crystal, it can be made almost totally transparent where maximum light passes through it. Another level of voltage will twist the molecules and practically block all light from being passed. In this way, the liquid crystal acts almost like a camera shutter but in this case with varying degrees of transparency being possible. The liquid crystal material is contained within glass plates. A bright light source such as a fluorescent lamp or high-brightness LEDs are placed behind the liquid crystal.

In front of the liquid crystal is another transparent sheet containing the red, blue, and green pixels described earlier. The whole thing is sandwiched together in a very thin screen. The electrical signals applied to a matrix of wires cause the liquid crystals to be properly oriented for each pixel and to allow the correct proportion of light through to each of the red, blue, and green light filters.

Plasma—A plasma screen is also made up of many tiny light sources. Each light source or subpixel is actually a small chamber containing xenon and neon gases. When a voltage is applied to each of these chambers, the gases ionize and emit ultraviolet light. The ultraviolet light then actually strikes a phosphor coated inside the chamber. The chamber then emits red, green, or blue light. The degree of ionization determines the brightness of each subpixel. In any case, electrical signals are then applied in sequence to the subpixels to turn them on with the correct degree of brightness to produce the desired color.

DLP—Digital light processing is a technique invented by Texas Instruments. The heart of it is an integrated circuit chip made up of hundreds of thousands or millions of tiny mirrors. The mirrors can be moved to one of two positions. This is an example of a *microelectromechanical system* (MEMS). MEMS is a technique for making mechanical moving objects using semiconductor manufacturing techniques.

Each of the tiny mirrors can be moved so that in one position it will reflect maximum light (representing white) and the other position where no light is reflected (representing black). You can then produce any shade of gray simply by switching the individual mirrors off and on at a high rate of speed. The light source is shined on the mirror and the resulting monochrome scene produced on the DLP chip is then focused on a screen from the rear by a lens system.

To produce color, the light source is passed through a rotating disc with alternating red, green, and blue segments. The speed of rotation and the frequency of each color on the disc allow each mirror to then reflect a specific color of light. In the newer DLP screens, high-intensity red, green, and blue LEDs are used to send light to the mirrors, which are then mixed and projected onto the screen.

LED—The newest TV screens are made of tiny red, green, and blue light emitting diodes arranged in a pixel matrix as described earlier. The LED screens are the brightest available and respond fast to action scenes. However, they are the most expensive screens.

3D TV

Yes, you actually can buy a 3D TV set that gives you that same 3D "coming-at-you" feeling of theater 3D. As this is written, these sets are very expensive and there is very little 3D material to see. Right now it is only available on DVD. Eventually it may be available by cable or satellite or over the Internet. And, yes, you still have to wear the special glasses. These glasses are not the red and green lens type but a special electronic version. The lenses use electrical polarization like a shutter that turns the left and right lenses off and on

Which Screen Type Is Best?

It is very difficult to find a new TV set with a CRT. You can still buy them, but they are typically only available in the smaller screen sizes such as 9, 13, and 19 inches.

LCD screens are available in a wide range of sizes, but the minimum today for a TV set is about 19 inches. Remember that these dimensions are the diagonal dimension from an upper corner of the screen to the lower diagonal corner. The most popular LCD screen size is 32 inches, but they are available in a wide range of sizes from about 42 to 65 inches. Even bigger ones are available. Plasma screens come in sizes from about 42 inches to about 70 inches. A DLP screen can be had in sizes from 50 to 100 inches.

As for prices, LCD and plasma screens are approximately comparable. DLP projection screens are significantly less expensive for the same size.

There are a couple of other factors to consider in evaluating screen types. As far as brightness goes, they are all very good, with the DLP probably slightly less bright than the other two types. Contrast, meaning the ratio of the brightest white to the darkest black, is also excellent on all types of screens, a little bit less so on the LCD.

Viewing angle means whether you can see the picture from any angle. Most screens are viewed directly straight on, but many viewers are off to the side or at an angle. DLP screens generally have poor viewing angles: when you get too far off to the side, you will not be able to see the picture or it will fade away. LCDs have this problem but to a lesser extent. Plasma screens have the widest viewing angles.

Finally, the ability of a screen to produce high-speed motion is generally excellent on plasma and DLP screens. LCDs did produce some smearing, but recently the higher scan rates of 120 and 240 frames per second have essentially eliminated this problem.

LCD screens dominate TV today and are gradually replacing all other types as their price continues to drop with better manufacturing methods. Plasmas and DLP are still available but are gradually fading. LED screens are growing in popularity and will become more widely used as prices come down in the future.

in synchronization with the two alternating sets of frames on the screen. A wireless signal (IR or radio) signals the glasses when to switch. The switching occurs at very high speed so the effect is to see the 3D scene. Only time will tell if this is just a gimmick or niche TV product.

CABLE TELEVISION

It is estimated that about 15% of all U.S. citizens get their TV over the air with an antenna. But over 70% get TV via cable. The remaining 15% get satellite TV. Why? Mainly because cable offers not only better, more reliable signals than by antenna, but also offers hundreds of channels of programming. But that's not all. Most of those who have cable also use the cable connection for

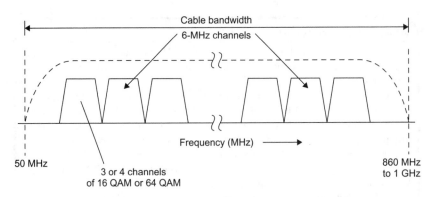

FIGURE 11.14   Coax cable bandwidth up to 1 GHz can contain about 130 6-MHz channels.

high-speed Internet access service. Of course, you pay for all that, and it has become like a monthly utility bill. With the cable industry being so big, it pays to know a little more about it.

It all starts with a facility called the *cable head end*. Another name you might hear is cable modem termination system (CMTS), which specifically is the equipment at the head end. This is a place where the company collects video and programming from many different sources. It snatches signals off the air, it receives premium programming like HBO, ESPN, Showtime, and Fox via satellites, stores movies on its large computers called video servers, and may even do some of its own local programming. The result is a hundred channels or more.

Now each of those TV signals—video plus audio—is modulated onto carriers that are spaced 6 MHz apart just like the regular over-the-air TV channels. Then all of those modulated carriers are added together to form a very complex signal. That signal is then amplified and put on the outgoing cable. Figure 11.14 shows what the spectrum looks like. Most cables have a bandwidth from about 50 MHz up to about 860 MHz to 1 GHz, which will accommodate up to about 130 6-MHz channels. We call this process frequency-division multiplexing, where a single cable can carry many separate channels of data, each on a different frequency and bandwidth. And since the cable is shielded, it does not radiate, so it, in effect, duplicates the free-space electromagnetic spectrum.

Since most TV today is digital, digital modulation methods are used to put multiple TV signals within each 6-MHz channel. Using 16-QAM or 64-QAM, a cable company can squeeze several channels of video/audio into each 6-MHz chunk of spectrum, providing hundreds of additional slots for programming. That is why cable companies can advertise up to 500 channels of entertainment, news, and so on.

All of the digital video/audio signals are then remodulated onto different channel assignments. All of these 6-MHz channels are added together or linearly mixed in a multiplexer (MUX) to create a massively complex composite signal. This is what goes out on the cable (see Figure 11.15).

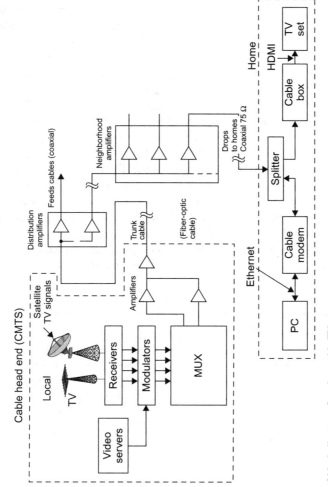

FIGURE 11.15 General block diagram of modern cable TV system.

The signal distribution network is called a *hybrid fiber cable* (HFC) system. This multiplexed signal is applied to a fiber optic cable, called a *trunk cable*. It is routed around the city and it terminates at selected points where the signal is divided into multiple paths of coax cable. These so-called *feeder cables* are connected to amplifiers along the way to overcome the large attenuation of the cables. At some point, the feeder cables come into your neighborhood and are further amplified and split into multiple signals. These signals are then sent to your house by a coax drop cable. The most common type is 75-ohm RG-6/U coax with F-connectors. The cable comes into your home and is usually split two or more times and sent to wall connectors in the various rooms. These connectors connect the signal to a cable box and/or a cable modem for the PC.

The cable box is like a tuner in that its main job is to select one of the many channels and send it to your TV set. Figure 11.16 shows a simplified block diagram of a typical cable box, usually called a set-top box (STB). The input is a TV tuner with channel selection via a frequency-synthesizer local oscillator by way of your remote control. The selected signal is then demodulated and decompressed by a digital media processor chip, and then sent to the TV set via one of the many possible interfaces, usually the HDMI or *high-definition multimedia interface*. This is a very fast digital interface with special cable and connector. Most TV sets have this as the most common input. Other video format interfaces such as RGB, PrPbPy, or S-video, are usually provided to accommodate older analog TV sets. An additional feature of many of the newer STBs is a built-in digital video recorder (DVR). This is essentially a computer hard disc drive that records the program you want to save for viewing later.

Other Wired TV Distribution Systems

There are a couple of other ways that digital video is distributed to homes. One is a fully fiber optic system. The fiber cables go all the way from the head end to the homes. This was once too expensive, but today, thanks to improved fiber and other technologies, it is cost effective. Such systems are generally referred to as *passive optical networks* (PONs). "Passive" simply means that the signals are strong enough and the cable runs short enough so that no amplification is needed along the way. The digital video and audio are modulated onto infrared (IR) light beams and sent down the cable. Because fiber has such wide bandwidth, data rates are extremely high, over 1 GB/second in the newer systems. This allows not only digital video to be transmitted but also very-high-speed Internet service. Such systems use standards like Ethernet PON or GPON for gigabit PON. A typical U.S. system is Verizon's FiOS.

Another new system is called IPTV for *Internet protocol TV*. This is video distribution via putting the compressed video on a fiber optic cable using the familiar TCP/IP protocols. The signals travel via fiber to the neighborhoods. From there they are split off and the signals travel via in-place twisted-pair telephone lines.

(Continued)

Other Wired TV Distribution Systems (Continued)

New forms of modems using a form of OFDM called *discrete multitone* (DMT) transmit the signal over the twisted pair at high speeds. These lines are now used just for high-speed Internet service and are referred to as *asynchronous digital subscriber lines* (ADSL or just DSL). Newer versions called ADSL2 and ADSL2+ or VDSL for video DSL provide data rates from 26 to 52 Mbps on standard phone lines. An example of a system in the United States is AT&T's U-verse system.

For Internet access, you connect a cable modem to the coax cable and it automatically selects the channel where the data services are assigned. The modem goes to your computer via an Ethernet cable. Access to the Internet is by way of your computer's browser.

SATELLITE TV

Satellite TV technically is also over-the-air TV or radio, but instead of the signals coming from one or more local TV stations, broadcast towers, the signals emanate from a geosynchronous satellite 22,300 miles away orbiting around the Equator (see Figure 11.17). The satellite transponders operate in the Ku band with frequencies of 10.95 to 14.5 GHz. This system is referred to as *direct broadcast satellite* (DBS).

The satellite TV company gets its TV like the cable companies, and brings it all together at a single Earth station head end where it is all collected and transmitted on an up-link to the satellite in the 14- to 14.5-GHz range (see Figure 11.18). The signals are then amplified in the satellite and rebroadcast on the down-link frequencies at 10.95 to 12.75 GHz. A high-gain antenna focuses the signals on the United States or other country of interest.

Down on Earth, satellite TV subscribers have a special receiver and antenna. These are illustrated in Figure 11.19. The antenna is usually a small dish or parabolic reflector antenna about 18 inches in diameter. The antenna is usually a small horn. Also inside the antenna assembly is a low-noise amplifier (LNA) to boost the small signal from the satellite. This is then sent to a mixer along with a local oscillator signal to down-convert the signals into the 950- to 2100-MHz range. This signal is then routed by coax to the satellite receiver. The LNA/mixer assembly at the antenna is called a low-noise block (LNB) converter. This receiver signal processing at the antenna is necessary because the Ku band frequencies of 10.95 to 12.75 GHz are so high they would be far too greatly attenuated by the antenna coax cable to the receiver. So down-converting to the lower frequency allows coax to be used to the receiver.

The satellite receiver is like the cable box in that it is used to select the desired channel and demodulate it for the TV set. The satellite signals are all digital and transmitted in packets 147 bytes long. A satellite receiver is shown in Figure 11.20. MPEG-2 video decompression is used. The recovered signals

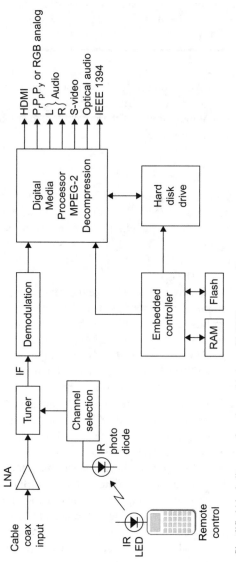

FIGURE 11.16 Simplified block diagram of modern cable box with DVR.

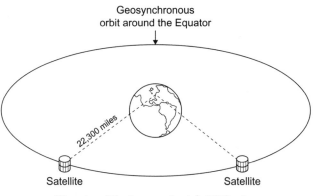

Geosynchronous
orbit around the Equator

22,300 miles

Satellite   Satellite

Diameter of Earth approximately 8000 miles

FIGURE 11.17   TV distribution satellites in geosynchronous orbits act as wireless relay stations.

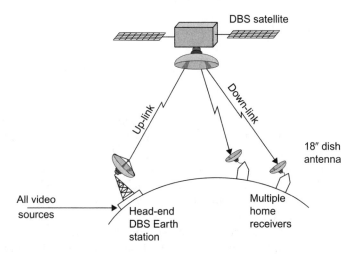

DBS satellite

Up-link   Down-link

18″ dish
antenna

All video
sources

Head-end
DBS Earth
station

Multiple
home
receivers

FIGURE 11.18   Video from all sources is transmitted up-link to satellite that rebroadcasts it back to Earth over down-link to home receivers.

are then sent to a digital TV receiver, typically via an HDMI port. Or in older systems, the video and audio are remodulated on a channel 3 or 4 TV carrier and sent to an analog TV set.

CELL PHONE TV

You knew it would come. Actual TV on your cell phone. You can easily get video via the Internet connection on a 3G phone over the network or via a Wi-Fi hot spot. Typical video is like that from YouTube or just video clips that users attach.

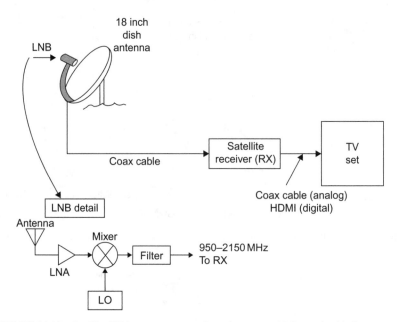

FIGURE 11.19 Satellite TV home system consists of antenna with low-noise block converter, receiver, and TV set.

Some cellular carriers also offer several video channels over the network for an extra fee each month. Such video is available from AT&T and Verizon, and consists mainly of news, weather, sports, and a few comedy channels.

Since video is so data intensive and requires lots of extra bandwidth and high speeds, too much video service can overload most cellular networks. It has been decided that the best way to deliver TV service to a cell phone is to broadcast it. Several methods are being implemented. For example, in Europe the DVB-H system is being used to broadcast programming to handsets. Similar systems in Japan and Korea are also used. Both are digital. In the United States, MediaFLO broadcasts video to handsets on the old TV channel 55 (716 to 722 MHz). It uses OFDM and about 15 to 20 channels are available. The service is provided for a monthly fee.

In the near future, standard ASTC digital TV will be available in the United States. A special version for mobile handsets has been developed and will likely be adopted by existing TV stations for broadcast to cell phones. This is expected to be free TV, but no doubt ads will be used to pay the bills. If making a phone call or texting by cell phone is distracting to drivers now, wait until TV is available. (Now that is something to look forward to.)

CLOSED-CIRCUIT TV

Closed-circuit TV (CCTV) is used primarily for surveillance. You have seen the small wall- or ceiling-mounted TV cameras in stores, airports, companies,

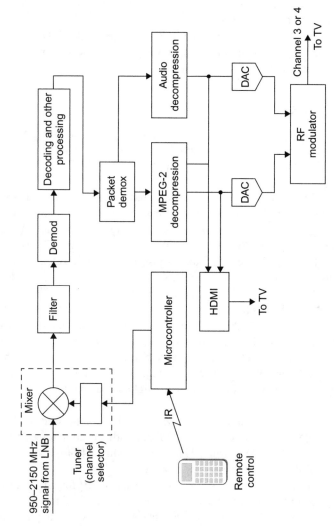

FIGURE 11.20 Simplified block diagram of satellite TV receiver.

and today on the streets. The video signals are routed to a central point where multiple TV sets or monitors are set up for observation.

In many installations, the video is transmitted directly over coax cable to the monitors. There is no modulation involved. However, in some cases, when remote cameras are used, a wireless link via 2.4 GHz is typical. Of course, modulation is then used in the transmission so that a receiver is needed at the viewing end to recover the video.

Many installations also have video recorders (VHS tape or DVD) to capture several minutes or hours of video. In the past, most cameras were black and white, but today more and more use color.

DIGITAL VIDEO DISCS

The formal meaning of DVD is *digital versatile disc*. Its main use is video but it can also store computer data or audio. It is the storage medium of choice for movies, music videos, and other video programming. It has virtually replaced the VCR.

The basic DVD storage medium is a disc that is almost identical to the CD discussed in Chapter 10. The disc is the same size (120-mm diameter), but it is formatted differently to store many more bits and bytes of data. While the CD can store up to about 700 MB, the DVD can store about 4.7 GB, that is, gigabytes or billions of bytes. And that is just the basic one-sided disc. There are also single-side/double-layer discs, double-sided/single-layer, and double-sided/double-layer versions that can store even more, up to 15.9 GB.

The storage method is the same. A laser beam burns pits into a spiral track. The data is represented as pits and lands as described earlier in the CD discussion (see Figure 10.6). The big difference is that the size of the pits is significantly smaller and the track spacing is less than half the CD spacing, that is, 1.6 microns (1600 nanometers) versus 740 nanometers (nm). A nanometer is 1-billionth of a meter. With smaller pits and tighter spacing, the disc can hold significantly more data.

But that is not the whole story. The video is also compressed. Unlike the audio on a CD, which is not compressed, digital video is compressed with the MPEG-2 standard described earlier. This gives roughly a 40-to-1 compression. In that way, a full 2 hours of video can be stored.

As for audio, the format is as described in Chapter 10 for CDs, but DVD audio is Digital Dolby Surround Sound 5.1 using six channels of 16-bit, 44.1-kHz digital sound. Audio DVDs are also available with 192-kHz sampling of the audio with 24-bit ADCs, giving even greater fidelity and low-noise performance.

How a DVD Player Works

Figure 11.21 depicts the block diagram of a typical DVD player. First, note the motor that drives the disc. It is a variable-speed device that changes speeds as the pickup mechanism moves from inside to outside as it reads the spiral track.

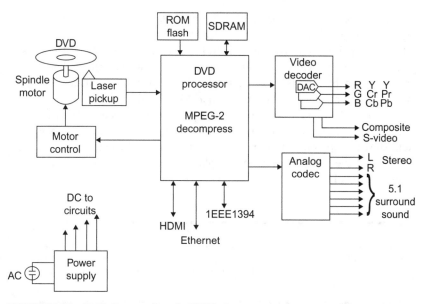

FIGURE 11.21   Block diagram of standard DVD player.

The pickup head uses a laser that shines on the disc to read the pits and lands. Because of the closer spacing, a higher-frequency (shorter wavelength) laser is used. CD players use a 780-nm laser, while DVD players use a 640-nm laser that has a smaller focused beam.

The photodetector in the head picks up the laser reflections and sends the 10 Mb/second data stream to the processing chips. In some players, a single-chip application-specific device handles all of the digital processing from demultiplexing, to error detection and correction, decoding, and MPEG-2 decompression. All this may be in separate chips in some players.

In any case, the MPEG-2 data goes to a video decoder and to multiple DACs that re-create the original analog RGB signals. Various video formats are used, such as RGB, YCrCb, and YPrPb, each of which needs three DACs each. Most DVD players also decode regular composite video and S-video.

The separated audio signals go to an audio codec that contains the DACs which recover the audio content either in basic stereo or 5.1 surround sound. As for interfaces, an HDMI output is provided. Most players also have an Ethernet port, IEEE 1394, and USB connections.

Don't forget that some of the later models are also DVD recorders. They play DVDs but also burn new ones with video from the TV set, cable box, satellite box, or a built-in tuner.

A Word about Blu-ray

Blu-ray is the next generation of DVD medium and players. The discs are the same size but thanks to even smaller pits, lands, and track spacing, even greater storage capacity is available. With 320-nm track spacing and a 405-nm (blue light wavelength) laser, the single-sided disc can hold up to 25 GB of data. That is enough to store two full-length high-definition movies or 13 hours of standard-definition video. A double-layer version is also available that doubles those storage figures. Blu-ray uses MPEG-2 compression, but also supports other video compression standards such as MPEG-4/H.264 and VC-1.

For a while, there were several other video disc standards competing for dominance. The most notable was HD-DVD, which did bring players to market. Most movie companies and other media sources selected Blu-ray as the distribution medium, making Blu-ray players the de facto standard and eliminating HD-DVD.

Project 11.1

Look at a Pixel
Get a magnifying glass and hold it up to your TV screen to see the individual color pixels or subpixels. Are they triads or rectangular areas?

Project 11.2

Diagram Your Own TV System
Draw a block diagram of your own home TV setup. Just draw a box for each and indicate how each of them is connected. Identify the type of cable, connector, and interface between each box.

Project 11.3

Try Over-the-Air TV
If you get your TV by cable or satellite, try receiving local channels on an antenna. Use an indoor antenna for simplicity. You can buy a set of "rabbit ears" from a local Radio Shack. Or try an outdoor antenna from Radio Shack. You can get some channels on a simple loop of wire 1 yard long.

If you still have an older analog TV, you probably also have a cable box and went through this process before.

Chapter 12

Industrial Control

How to Automate: Monitor, Process, and Control

In this Chapter:
- Control defined.
- Open and closed control systems.
- Sensors and transducers.
- Actuators like relays, solenoids, and motors.
- Thyristors.
- Programmable logic controllers.

INTRODUCTION

Industrial control is one of the four major applications of electronics. In industrial control, electronic components, circuits, and equipment are used to operate various types of machines in manufacturing plants. Most industrial machines like robots are mechanical in nature but electrically operated. Most are powered by electric motors or other actuators. To operate the machine, electrical power must be properly controlled. Industrial control electronics is used to turn machines off and on at the appropriate time, control their speed of operation, and otherwise produce the desired manipulations.

Electronic control is not confined to industry. There is lots going on in the home and car as well. The principles given here apply to those applications as well.

OPEN- AND CLOSED-LOOP CONTROL

The basic industrial control process is illustrated in Figure 12.1. Electrical inputs, such as switches, sensors, or other devices, are used to initiate an electronic control process. The signal-processing circuits generate control output signals that are used to operate the industrial machines. This basic process is known as open-loop control. The typical device being controlled by the output signals is an electric motor that operates a machine. Other outputs could operate lights, relays, solenoids, or a variety of other devices.

doi: 10.1016/B978-1-85617-700-9.00012-6

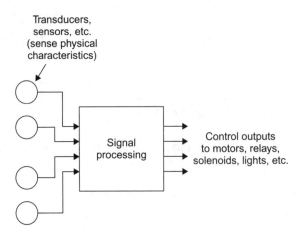

FIGURE 12.1 General block diagram of industrial control process.

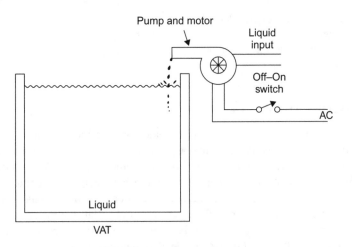

FIGURE 12.2 Open-loop control system for filling vat with liquid via a pump.

Open-Loop Control

A simple example of an open-loop control system is shown in Figure 12.2. The input is a switch that applies electrical power to a motor. The motor, in turn, operates a pump that causes liquid to be put into a vat or tank. As soon as the vat is full, the operator turns the switch off. The basic control process in this system is simply turning power off and on to control the pump. The operator visually monitors the level in the vat until the correct level is reached.

Closed-Loop Control

An improved form of industrial control system is illustrated in Figure 12.3. It is known as a closed-loop control system. The system is given an initial input

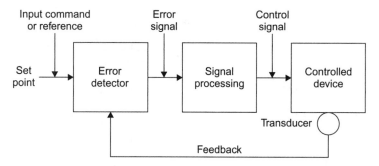

FIGURE 12.3   Generic diagram of any closed-loop control system with feedback.

reference called the set point. The set point is usually a voltage that represents some physical value that is to be achieved. It is applied to an error detector that compares this input to a signal from the controlled devices. If the two signals are different, an error output is produced. This error signal is processed (usually amplified), and an output control signal is derived from it. It is this output signal that operates the controlled device. A transducer monitors the controlled device to see that the desired outcome is obtained. The signal from this transducer is called *feedback*. The feedback signal is compared to the input to produce the error signal. The control signal operates the controlled device until the error is reduced to zero. At that time, the desired output condition, where the feedback equals the set point, is obtained.

The key to closed-loop control is the feedback. Feedback signals tell the system whether the machine is performing correctly. The feedback signal allows the system to adjust itself in such a way that the desired output is continuously accomplished. Control is automatic.

A closed-loop version of our system to fill a vat with liquid is shown in Figure 12.4. An initial input signal is applied to start the process. If the vat is not full, the transducer will sense it and send a signal to the error detector that says the liquid level is low. The error detector generates an output signal that is processed and used to start the pump motor. The motor rotates the pump, which puts liquid in the tank. When the vat is full, the sensor will generate a signal that is sent to the error detector. The error is now zero, so the pump motor is turned off.

The primary benefit of a closed-loop control system is that its operation is automatic. In the open-loop system, an operator has to turn the system off and on manually to fill the tank. In the closed-loop control system, the operator gives an initial input, called the set point, to which the error detector will compare the feedback signal from the sensor. He or she turns the system on. From that point on, operation is automatic. The pump turns itself off when the vat is full. If the liquid level drops below the desired level due to usage, the sensor indicates a low level and automatically turns on the motor so that the pump will again fill the tank. When the tank is full, the unit shuts off by itself.

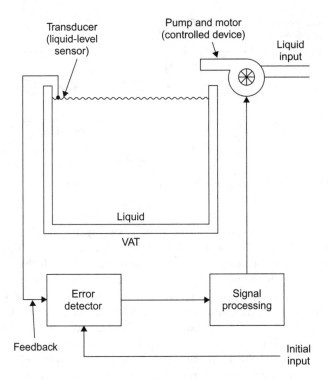

FIGURE 12.4   Filling vat with liquid automatically with closed-loop control system.

Most industrial control systems are closed loop to make them automatic. And your home has some as well. Your toilet flusher is a closed loop as is your heating and air conditioning (HVAC) thermostat. Appliances like your washer, dryer, and dishwasher are other examples of closed-loop control. All of them contain a microcontroller with sensors, timers, and feedback to make them totally automatic.

Controllers

While many industrial control applications are implemented with simple control circuits, most industrial control is carried out by specialized instruments called controllers. These units accept inputs from sensors, condition them, process them, and then generate output control signals.

A good example of an analog controller is one used to control the temperature in a system. A resistive temperature sensor is used to monitor the temperature of the liquid in a pipe. If the temperature goes up, the sensor output goes up. This is sensed by the controller circuitry. Inside the controller, a comparator is set to a desired temperature threshold. Should the temperature exceed a specified set point, the controller will turn off the device heating the

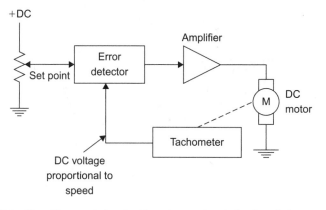

FIGURE 12.5   Closed-loop control system for motor speed control and regulation.

pipe and liquid. Should the temperature drop below another set point temperature, the controller will turn on the heating device in an effort to maintain the
set point temperature. While the heating element will be turning off and on, it
will produce the desired average overall temperature.

A continuous analog controller will provide proportional control rather than
an off/off response. For instance, suppose that you want to control the speed of
a motor. The system depicted in Figure 12.5 does this. You want the motor to
stay set to the speed you want, say 1200 rpm. A pot connected to a voltage represents that speed value. The voltage is applied to an error detector along with the
feedback input from a tachometer. A tachometer is a device that senses rotational
speed and develops a proportional DC voltage. That voltage is sent to the error
detector and compared to the set point. If they are different, an error voltage is
developed. This voltage is then amplified and applied to the motor. The speed of a
DC motor is proportional to the voltage applied to it. So if the motor slows down,
the error voltage is increased and that is amplified and applied to the motor speeding it up to compensate for the speed drop. A speed up in the motor produces an
increased tachometer output. The resulting error signal causes the motor speed to
drop until the error is zero. With such a circuit, the motor speed remains constant.

An analog controller is a product made up of the set point pot, error detector, and a selection of amplifiers whose gain can be varied and an output
amplifier that can drive the motor or other device. Many different types of
commercial electronic controllers have been developed to monitor and control
physical characteristics such as pressure, physical strain, weight, liquid flow
rate, physical position, or liquid level in a tank. Digital controllers that contain
a microcomputer are also available.

SENSORS

An essential element of most control systems is the inputs from sensors. Sensors
are the components that detect physical changes or events and convert them into

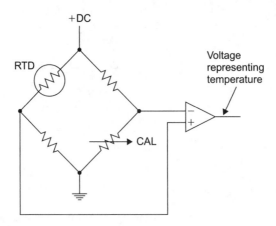

FIGURE 12.6 Most resistive sensors like an RTD or thermistor are placed in a bridge circuit to condition them to provide amplified output voltage.

electrical signals to be processed. Sometimes the term transducer is used to refer to the sensor. A transducer is a device that converts one type of energy into another such as mechanical energy to electrical. There are literally hundreds of different types of sensors. A few examples of the most common types follow.

Temperature Sensors

Temperature is probably the most often sensed physical characteristic. Temperature sensors convert temperature to a resistance change or voltage change. The most commonly used temperature sensors are RTDs, thermistors, and thermocouples.

RTD—An RTD is a resistive temperature device. It is essentially just a platinum wire whose resistance varies with temperature. Most of them have a resistance of 100 ohms at zero degrees Celsius (0°C). As the temperature changes, the resistance changes to reflect the change. RTDs have what we call a linear positive temperature coefficient. That is, as the temperature goes up the resistance goes up and vice versa.

To convert the resistance variation into a voltage, the RTD is usually put into a bridge circuit as shown in Figure 12.6. The bridge is initially balanced with the variable calibrate control (CAL) to produce zero output at a given temperature. Then if the temperature varies, the bridge becomes unbalanced and a voltage is produced. This is then amplified into a larger voltage that is a measure of the temperature. The output can be measured with a meter or it can be digitized for use as in input to a digital system for display or control.

Thermistor—A thermistor is a resistor whose resistance has a negative temperature coefficient. As temperature goes up the resistance goes down and vice versa (see Figure 12.7A). The resistance change is much greater than that

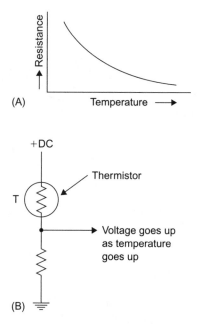

FIGURE 12.7 Characteristic curve of thermistor and simple voltage divider circuit producing output voltage.

of an RTD. The thermistor can be used in a simple voltage divider to develop an output voltage or in a bridge circuit like the RTD (see Figure 12.7B).

Thermocouple—A thermocouple is a unique type of temperature sensor, as it develops a voltage rather than a resistance change. It is formed with two dissimilar metals. If the junction of the metals is heated, a voltage is developed. The voltage is usually in the millivolt range so that it usually has to be amplified before it becomes useful. The thermocouple's main advantage is its accuracy at very high temperatures.

Solid State—Diodes and transistors make good temperature sensors. A silicon diode has a very linear voltage drop variation that decreases with an increase in temperature. This is true of the emitter-base junction of a bipolar transistor as well. Zener regulator diodes exhibit a similar response. The sensor is usually packaged into a complete integrated circuit with an amplifier that produces a voltage output proportional to the temperature variation.

Pressure Sensors

A pressure sensor responds to force or pressure. There are many different types. A widely used kind is called a *strain gauge*. A strain gauge is essentially a thin pattern of metal deposited on a plastic base like that shown in Figure 12.8. It has a specific value of resistance. A value of 120 ohms is common. The strain

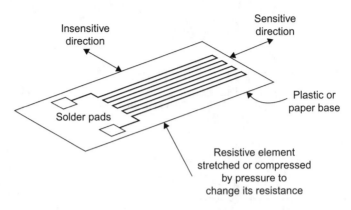

FIGURE 12.8   Strain gauge is a resistive element whose value changes with stress or pressure.

gauge is then glued or cemented with an adhesive to the object to which the pressure is to be applied. An example is a steel beam. When pressure is applied to the beam, it will bend. As it bends, it stretches or compresses the strain gauge resistance element. If the resistive element is stretched, its resistance increases. If it is compressed the resistance decreases. Knowing the specific character-istics of the resistance change per pound of pressure, an exact measure of the pressure can be determined. Mostly strain gauges are used in a bridge circuit like that shown in Figure 12.6. Since the resistance variation is small and the resulting bridge circuit output is small, a special amplifier called an *instrumen-tation amplifier* is used. It has differential inputs and very high gain.

A special pressure sensor called a *load cell* is typically made up of strain gauges. It is fully calibrated to provide a specific voltage output with a specific pressure. Load cells are widely used for weight measurement. An example application is the use of load cells in truck weighing scales.

There are other forms of pressure transducers. Many are solid-state devices with an output voltage proportional to pressure. *Piezoelectric pressure trans-ducers* use the voltage produced by certain types of crystals or ceramics when pressure is applied to them.

Switches

Switches are widely used as sensors. Switches are mechanical devices with electrical contacts that open or close with a particular physical movement. Switches are used to detect mechanical movement and specifically to deter-mine when the limit of some physical device has been reached. Called *limit switches*, these devices have various arms and mechanisms that permit ease of use in a variety of situations. Examples of such switches are those that can detect when a cabinet door is open or closed or a switch that senses when a mechanical device reaches a specific point where motion must be stopped.

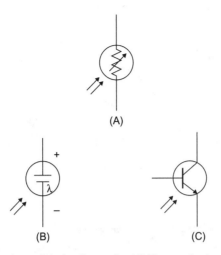

(A)

(B)   (C)

FIGURE 12.9   Schematic symbols for photo cells. (**A**) Photoconductive or photoresistive. (**B**) Photovoltaic or solar cell. (**C**) Photo transistor.

And of course, a human-operated switch may be the most common. It is usually a button that is pushed to initiate or stop some action.

Potentiometers

A potentiometer is a variable resistor or variable voltage divider. While most electronic pots are rotary devices, they also come in a linear variety. The variable arm of the pot is controlled by some external mechanical device. Then by using the resistance value of the pot or the voltage it produces as a variable voltage divider, it can be used to sense mechanical position.

Photoelectric Sensors

One of the most widely used industrial transducers is a photoelectric sensor. A photoelectric sensor is a component that responds to light and produces an electrical signal that can initiate some operation.

There are three basic types of photoelectric sensors. These are the *photoresistive* or *photoconductive cell*, the *photovoltaic cell*, and the *phototransistor*. The photoconductive sensor is a light-sensitive resistor. Its schematic symbol is shown in Figure 12.9A. The resistance of the device varies with the amount of light falling on it. With no light falling on the sensor, its resistance will be very high, 100,000 ohms or more. When bright light shines on the sensor, however, its resistance will drop to a very low value, usually several hundred ohms or less. Of course, at a light level between bright and dark, the resistance of the device will be somewhere between 100 ohms and several hundred thousand ohms. The resistance is inversely proportional to the light level.

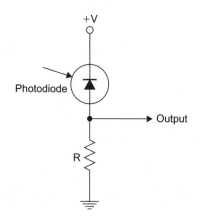

FIGURE 12.10   Photodiode and its simple output circuit.

Another type of photoelectric device is the photovoltaic cell, whose schematic symbol is illustrated in Figure 12.9B. This is the solar cell described in an earlier chapter. Whenever bright light falls on the cell, it will generate a small DC voltage. The maximum output of a typical photovoltaic cell is approximately 0.45 to 0.5 volt. At lower light levels, the output voltage will be less. Photovoltaic cells are used primarily in power generation systems for producing charging voltage for batteries in satellites and other remote systems. Multiple cells are connected in various series and parallel combinations to get the desired voltage level and current capacity. In some applications, the photovoltaic cell can be used for industrial control. These are also found in portable calculators. Two AA cells at 1.5 volts each produce 3 volts in series. To generate 3 volts with 0.5-volt solar cells, you would need to connect six in series ($6 \times 0.5 = 3$ volts).

A silicon diode can also be used as a light detector. Special diodes in clear housings are often used to detect light. These are called photodiodes. Photodiodes are operated in the reverse-biased direction, as shown in Figure 12.10. With no light falling on the diode, it is cut off, so there is no voltage across R_1. When light strikes the diode, its reverse leakage current increases dramatically, causing a voltage drop to occur across R_1.

A phototransistor is a light-sensitive transistor. Its symbol is given in Figure 12.9C. It only has two terminals, the emitter and the collector. The device has a base, but typically it has no lead. Instead, the device is built so that when the base is exposed to light, it will cause the emitter-base junction to conduct. When light strikes the base, it causes ionization, which simulates base-current flow. Therefore, the transistor conducts between emitter and collector. The primary advantage of a phototransistor is its very high sensitivity over a photodiode. It will conduct even with a small amount of light applied to the base. The transistor essentially provides amplification so that small light levels can control a large current.

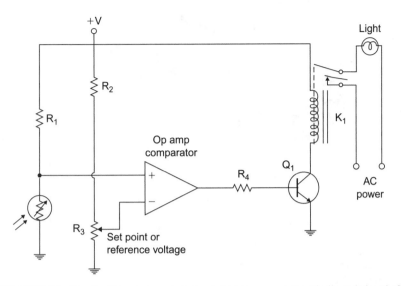

FIGURE 12.11 Simple off/on photo control circuit. Light turns on automatically at dark and off when light is present.

A typical closed-loop control circuit using a photoelectric cell is illustrated in Figure 12.11. This circuit will automatically turn a light on when it gets dark and turn the light off when it is light again. Note that a photoresistive cell is connected as part of a voltage divider with resistor R_1. The DC voltage from this voltage divider is applied to one input of an op amp comparator. The DC reference input to the comparator comes from a voltage divider made up of R_2 and R_3. The potentiometer can be used to adjust the DC reference voltage or set point. The output of the comparator drives a switching transistor Q_1, which will turn the relay K_1 off or on. This, in turn, operates the light.

When the photoresistive cell is exposed to light, its resistance will be very low. Therefore, the voltage from the voltage divider will be low. It will be less than the DC set point voltage and the comparator will not be triggered. The output of the comparator will be a negative voltage, which will reverse bias the emitter-base junction of the transistor and keep it turned off. As soon as it gets dark outside, the resistance of the photocell will increase. This will cause the voltage-divider output to increase. At some point, the voltage will be higher than the set point voltage, causing the comparator to switch. This applies a positive voltage to the transistor base, turning it on, and then turning on the relay and the light. The light will remain on as long as the photoconductive cell is in the dark. But as soon as light shines on the photocell, the voltage-divider output voltage will drop, thereby triggering the comparator and turning off the transistor, the relay, and the light bulb. By adjusting the reference or set point voltage with potentiometer R_3, the triggering point can be varied, so that the light will be turned off or on at the desired level of ambient lighting.

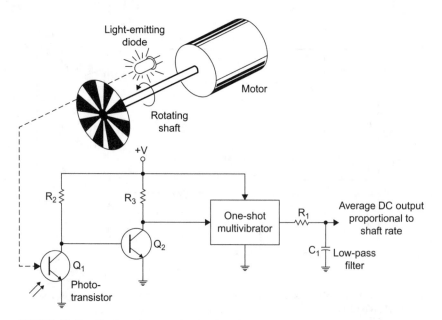

FIGURE 12.12 Simple tachometer circuit using phototransistor to create pulses that are averaged into a DC voltage proportional to speed.

This same basic concept can be used to produce a variety of industrial control operations. For example, you could use the circuit as an entry or intrusion detector. Shine a light on the photocell. If someone or something comes between the light and photocell, the circuit will trigger and the relay could then be used to operate a bell or buzzer or light.

Another example of a photoelectric controller is a tachometer circuit used for measuring the speed of a rotating shaft. Figure 12.12 shows a small, plastic wheel connected to the shaft of a motor. There are alternate clear and opaque areas on the disk. The disk is positioned so that it is between the light-emitting diode (LED) and the phototransistor. As the disk rotates, the dark opaque areas will block the light, while the clear areas pass the light from the LED to the phototransistor. When a clear area on the disk passes, the LED light will shine through it, hitting the phototransistor, and turning it on. When a dark area passes, the light will be blocked and the phototransistor will turn off. The rapidly rotating disk will cause the transistors to turn off and on very rapidly. This rapid conduction and nonconduction of the phototransistor causes a series of pulses to occur at the output of Q_2. These pulses are used to trigger the one-shot multivibrator.

The one-shot multivibrator produces a series of fixed, amplitude-width pulses. These are applied to an RC low-pass filter made up of C_1 and R_1. This filter will average the pulses into a DC voltage. This output voltage is proportional to the motor speed. If the speed of the motor should increase, the pulses occur at a more rapid rate. This means that the pulses occur closer together, and thus the average voltage across C_1 goes up. A DC voltmeter connected to the output can be used to indicate the speed. At low speeds, the meter indication will be low, while at high speeds it will be high. The meter dial itself can be calibrated in terms of revolutions per minute (rpm), thereby giving an accurate speed indication.

Flow Sensors

These devices have liquid or gas pass through them and measure the amount of flow in gallons per minute or other measure. They develop an output voltage proportional to flow rate.

There too many other sensors to catalog here but in most cases there is a sensor to convert any physical variable into a voltage to be measured.

OUTPUT DEVICES

An output device or an *actuator* is the thing that is most often controlled in a control system. The two most common types of actuators are relays and motors. There are many types of both.

Relays

A relay is an electromagnetically operated switch. A typical relay, and its schematic symbol, is illustrated in Figure 12.13. Its main components are a magnetic coil and a set of switching contacts. With no current applied to the coil, the spring keeps the armature pulled down, so that the contact arm makes connection with the upper, normally closed (NC) contact. When current is applied to the coil, a magnetic field is produced. This magnetic field attracts the steel armature, opening the upper contact, and closing the lower, normally open (NO) contact. The contacts form a single-pole, double-throw (SPDT) switch. When power is disconnected from the coil, the magnetic field ceases, and the contacts return to their initial state because of the spring.

Relay coils are designed to be operated from either AC or DC. The contacts are small circular disks of silver or tungsten. These materials are chosen because they will handle a large amount of current with minimum burning, arcing, and pitting.

The various kinds of switching contacts used in relays are illustrated in Figure 12.14. Normally open (NO) contacts are referred to as *form A*. Applying power to the coil closes the contacts. Normally closed (NC) contacts are referred to as *form B*. When power is applied to the relay coil, the contacts open. Single-pole, double-throw switching contacts are referred to as *form C*. When power is applied to the relay coil, the contact arm moves from contact A to contact B. Form C contacts are usually of the break-before-make variety. That is, contact A opens before the arm touches or "makes" contact B.

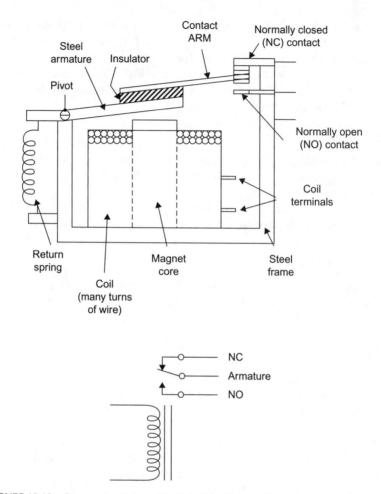

FIGURE 12.13   Cross-sectional view of typical relay showing major components and schematic diagram.

Sometimes you will hear the expression that a relay has been "picked" or "picked up." This means that power has been applied to the relay coil, and its contacts have been moved according to their function, and switching of a circuit has occurred.

Most relays usually have more than one set of contacts, so that multiple switching operations can be performed simultaneously. The various contact arms are ganged together so that they operate together when power is applied to the coil. A common configuration is two form-A contacts ganged to form a double-pole, single-throw (DPST) relay. Putting two form-C contacts together on the same relay produces a double-pole, double-throw (DPDT) relay.

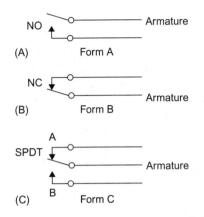

FIGURE 12.14   Common relay contacts. **(A)** Form A. **(B)** Form B. **(C)** Form C.

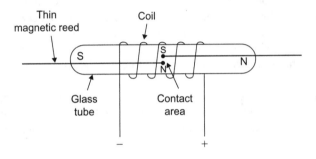

FIGURE 12.15   Reed relay.

Another widely used type of relay is the *reed relay*, illustrated in Figure 12.15. The heart of the reed relay is a reed switch that consists of two thin, metallic, reed contacts sealed in a glass tube. The reed contacts are made of a magnetic material, so that they will be magnetized when a magnetic field is applied to them. A coil consisting of many turns of fine wire is wound on a form and placed over the reed switch. When power is applied to the coil, a magnetic field is produced. This magnetic field magnetizes the two reeds as if they were bar magnets. One end of each reed will have a north pole and a south pole. The reeds are positioned so that the north pole of one will attract the south pole of the other. Remember, magnetic theory says that unlike poles attract, while like poles repel. The result is that the contacts will move toward one another and touch, making a good electrical connection. When power is removed from the coil, the contacts become demagnetized and spring apart. The reed relay will also operate by moving a permanent magnet near it. When the magnet is near, the contacts will close. This is a common way to sense open windows or doors in a home security system.

The contacts on reed relays cannot handle as much power as the contacts on conventional relays because they are smaller. As a result, they are used only in lower current power applications. However, the contacts on a reed relay open and close faster than those of a larger standard relay. Reed relay contacts can open and close in a few hundred microseconds, while standard relays require many milliseconds to open or close because of the larger contacts and the related structures that must be moved.

Relays are used in two basic ways in industrial control:

1. To control a larger current with a smaller current.
2. For remote control.

Let's examine the use of a relay for controlling a larger current with a smaller current. A good example of this is the starting system in your car. A large, heavy-duty DC motor is used to spin the flywheel on the car's engine to start it. This motor draws a current of hundreds of amperes. To turn the engine and start the car, the battery must be connected to the starter motor. This could be done directly by having the ignition switch connect the battery to the motor. However, this switch would have to have very large contacts to handle the high current. Such a switch would be large and expensive to build. The large heavy wires from the battery to the motor would have to be lengthened and brought up into the car.

A better solution is to use a smaller ignition switch to control the current applied to a relay coil (see Figure 12.16). The relay coil draws considerably less current than the motor; therefore, the switch contacts need not be large

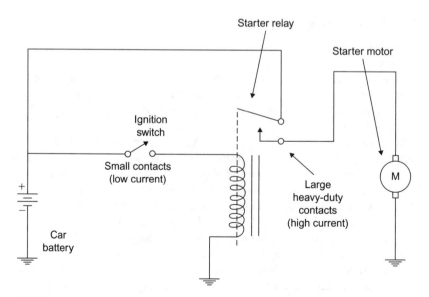

FIGURE 12.16 Controlling a large current with a smaller current with a relay.

or expensive. The switching contacts on the relay are much larger and will carry the high motor current. When the ignition switch is turned, its contacts close, applying current to the relay coil. The form-A contacts close, connecting the battery to the starting motor. When the car starts, the ignition switch is released, the relay drops out, and the starter motor shuts off.

Another application of a relay is remote control. Again, the starting system of a car is a good example. The load to be controlled, in this case the starter motor, is located remotely from the ignition switch. If the ignition switch were used to apply current to the motor, long wires would be required to connect the battery, the motor, and the switch. Heavy wire conductors would have to be used to carry the high current. The longer the interconnecting wires, the higher the resistance of the circuit. While large conductors have minimum resistance, when high current flows through them, voltage drops still occur. The voltage dropped across the interconnecting lines subtract from the battery voltage available for the motor. The voltage drops might be excessive, unless very large, expensive, and inconvenient conductors are used.

To eliminate this problem, a remote relay is used. The relay itself is located near the starter motor and the battery. In this way, the interconnecting leads can be kept short, thereby minimizing voltage drops. Since the relay operates from very low current, longer interconnecting wires can be used to attach it to the ignition switch without noticeable voltage drops.

In some industrial control applications, the load to be operated is located at extreme distances from the switch used to control it. The current controlling the device does not have to be passed through very long conductors if a relay is used. Small conductors running to the switch remotely from the load control the relay while the relay turns the load off and on.

Relay contacts stay closed (or open) as long as power is applied to the coil. This means that the switch operating the relay coil must be kept closed if the relay is to remain actuated. There are applications, however, where it is desirable to use a momentary contact switch to operate the relay. An example is a pushbutton that, when depressed, causes contacts to be closed and power to be applied to the relay. When the button is released, a spring causes the contacts to open. With such an arrangement, the relay will close momentarily while the button is depressed, but will drop out when the button is released. The objective is to cause the relay to latch into its actuated position when the coil is momentarily pulsed. Such a relay is called a *latching relay*. Some latching relays use a mechanical arrangement on the armature to cause the contacts to latch closed when the coil is momentarily pulsed.

A relay can be made to latch electrically if it has an extra set of contacts. A typical latching circuit is shown in Figure 12.17. This relay has two sets of form-C contacts that operate simultaneously. The lower set of contacts is used to operate the load. The upper set of contacts is used to automatically latch the relay once it is pulsed. Switch S_1 is a normally open (NO) momentary contact pushbutton that is used to apply power to the relay coil. Depressing the button

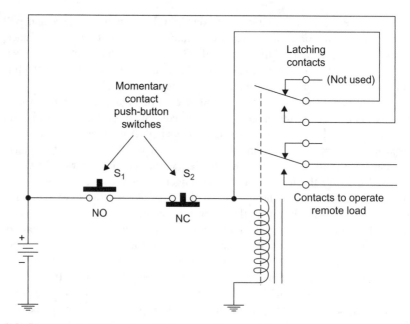

FIGURE 12.17   Latching a relay with its own contacts.

causes the relay coil to operate, closing the contacts. The lower set of contacts operates the load as desired. The upper set of contacts applies the power to the coil. Thus, when pushbutton S_1 is released, the circuit to the relay coil remains closed through the upper-latching contacts. To turn off the circuit, normally closed (NC) pushbutton S_2 must be depressed. This breaks the relay coil circuit, removing power. The relay drops out, opening the load contacts and the latching contacts.

While many relay coils are operated by mechanical switches, others are operated by a transistor. A typical transistor-controlled relay circuit is shown in Figure 12.18. The transistor is operated as a switch. When the base is grounded through contact B on S_1, the transistor is cut off and acts as an open switch. Therefore, no current flows in the relay coil. When a voltage is applied to the base resistor, through contact A on S_1, the transistor saturates, acting like a very-low-resistance or "on" switch. Current flows through the relay coil, operating the contacts. Removing the base current from the transistor turns the transistor off, terminating current in the relay. With this arrangement, an even smaller current can be used to control the already small relay current. A base current, of only microamperes, is sufficient to cause the transistor to operate.

In the circuit of Figure 12.18, when base current is removed from the transistor, the transistor will cut off. When current through the relay coil ceases, the magnetic field around it collapses. In doing so, the magnetic field induces a very high voltage in the relay coil. This very high voltage will be applied to the collector of the transistor. This voltage could be as high as several hundred, or

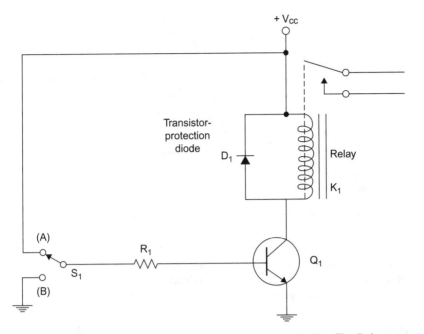

FIGURE 12.18   Operating a relay with a transistor is a common application. The diode protects the transistor when power is removed from the relay coil.

even thousands, of volts, depending on the relay coil. Such a high voltage will damage the transistor. To protect the transistor against such occurrences, some kind of voltage-spike eliminator or surge suppressor must be used across the relay coil. A commonly used protection device is a silicon diode D_1, as shown in Figure 12.18. When current is applied to the relay coil through the transistor, the diode is reverse biased and has absolutely no effect on the circuit. When the transistor turns off, the high voltage will be inducted into the coil with the polarity reversed from the supply voltage V_{cc}. This causes the diode to become forward biased and conduct. Thus, the diode effectively shorts out the high voltage spike, protecting the transistor.

Solenoids

A solenoid is a device that produces linear motion with an electromagnetic coil (see Figure 12.19). A magnetized steel rod called a plunger is inserted into a coil of wire. When a voltage is applied to the coil, a magnetic field is produced. It interacts with the magnetic rod producing attraction and/or repulsion that causes the rod to move into or out of the coil. The rod is usually physically constrained in one or both directions, and is often spring loaded so that it always returns to the same place when power is removed.

Solenoids produce linear motion. They are usually connected to a lever or ratchet mechanism that produces other forms of motion including rotation.

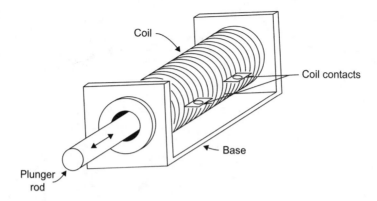

FIGURE 12.19 A solenoid produces linear motion using magnetic principles.

Solenoids are also used in valves to turn off the flow of a liquid or gas. By continuously controlling the current in the coil, the rod may also be precisely set to some position other than off or on. This effect is used in valves to continuously control the opening.

Motors

A motor is a device that converts electrical energy to mechanical energy. The mechanical output is typically the rotation of a shaft. That rotation can then be used directly or converted into other mechanical forms with levers, gears, ratchets, or cams. There are dozens of different motor types but most are categorized as either DC or AC.

Motors: Billions and Billions Sold and Used

Motors are one of those devices we take for granted. We use them every day but don't realize it. Most motors are hidden but do an enormous amount of physical manipulation. In industry, motors drive machine tools, conveyer belts, pumps, and other mechanical systems. Elevators use big motors.

At home motors are everywhere. Just think a minute and try to identify all the motors you have in your home and in your car. All of your appliances have motors, such as the washer, dryer, dishwasher, and HVAC units. There are motors in your mixers, blenders, and can openers. Shavers and hair dryers have motors. An analog clock is a motor. Then there are the motors in your electronics. In a PC or laptop, there are the fan motor and the ones in your hard disk and CD drives. Your printer, scanner, and copier and fax machine have motors. Your DVD player, CD player, and VCR all have multiple motors.

Then in your car is the starter motor, as well as the motors for windows, seats, mirrors, and windshield wipers.

Can you name some others?

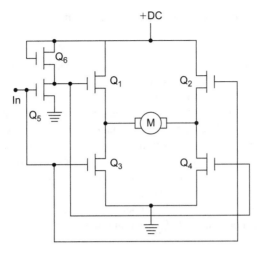

FIGURE 12.20   Using an H-bridge circuit to reverse the current flow and direction of rotation in a DC motor.

DC Motors

A DC motor operates from a DC voltage source like a battery or power supply. It is made up of a magnet and a rotating coil. When a current is applied to the coil, it forms an electromagnet. The magnetism of the coil interacts with the magnet field, producing rotary motion. The magnet is commonly a permanent magnet in smaller DC motors but an electromagnet in larger motors. In the larger motors, a special field coil is energized with DC voltage as well to create the magnetic field.

The main specifications of a motor are its horsepower rating, speed, and torque. Speed and torque are controlled by the amount of current flowing in the coil and/or the strength of the magnetic field. Increasing the voltage applied to the motor usually causes speed and torque to increase. Some kind of electronic control circuit is usually involved to set the speed to the desired value. A variable resistor in series with the motor will control its speed, but usually more complex electronic circuits are involved.

Incidentally, the direction of rotation is determined by the polarity of the DC voltage. Reversing the polarity reverses the magnetic field, thus changing the direction of rotation. Figure 12.20 shows a neat circuit used to change the direction of rotation of a DC motor. This is called an H-bridge and is made up of MOSFET switches. Q_2 and Q_3 are turned on by a positive input voltage IN. Q_5 inverts this voltage to zero, which keeps Q_1 and Q_4 off. Therefore, current (electrons) flows from ground through Q_3 through the motor (M) from left to right through Q_2. Setting the input to zero turns off Q_2 and Q_3 and turns on Q_1 and Q_4. Now current flows from ground through Q_4 and the motor from right to left and then through Q_1. This changes the direction of the motor with just a simple off/off input control voltage.

Most DC motors rotate too fast for many applications. A typical speed range is 1000 to 10,000 rpm. To achieve lower speeds, the motor is used in conjunction with a gear box or pulleys.

A special type of DC motor is the *stepper motor*. This is a DC motor driven by two DC pulse signals. The pulses operate coils or windings that produce a rotating magnetic field. A permanent magnet rotor interacts with the rotating magnetic field to produce motion. The interesting characteristic is that the rotation occurs in increments or steps instead of a continuous rotation as with a standard DC motor. Depending on the number of windings and coils and elements on the rotor, the increments usually vary in increments of 22.5, 18, 15, 7.5, or 3 degrees or smaller. The stepper motor therefore permits very precise positioning. Yet it can be rotated at several thousand rpm if needed.

AC Motors

AC motors usually operate from a 50- or 60-Hz power line. Again the rotation is produced by interacting magnetic fields. The speed of rotation is fixed by the number of coils and magnet poles inside the motor. That speed is also often higher than needed, so gears and/or pulleys are used to drop the speed to the desired value.

To change the speed of an AC motor, you must vary the frequency of the AC voltage applied to it. This means that you cannot operate the motor directly from the AC power line if the speed must be changed. Instead, you must use a so-called AC drive. This is a box of electronics that generates a variable frequency AC voltage to apply to the motor. An AC drive takes the AC line input and uses semiconductor switches to produce an AC signal at a different frequency.

Other Actuators

Besides motors, solenoids, and relays, there are a variety of different controlled devices. Valves are common device. Valves are used to turn the flow of a liquid or gas on or off. Proportional valves very carefully regulate the amount of flow.

Other devices are hydraulic or pneumatic devices. Hydraulic devices use oil in cylinders to provide mechanical force where needed or position something mechanically. Pneumatic cylinders work with compressed air to apply force or control the position of a device.

THYRISTORS

Many industrial switching circuits are implemented with MOSFETs designed for high voltage and high power. You will sometimes hear these devices called *insulated gate field-effect transistors* (IGFETs).

Other power-switching devices are also widely used, especially thyristors. A thyristor is a semiconductor device used for switching purposes in industrial control. Like the relay, it is used to apply and control electrical power to motors, heating elements, lights, and other loads in industrial applications. In many applications, thyristors replace relays. A thyristor acts as a switch, but because of its solid-state nature, is far faster in switching than a relay. There are three basic types of thyristors: silicon-controlled rectifiers (SCRs), triacs, and diacs.

Silicon-Controlled Rectifier

The SCR is a three-terminal thyristor that acts like a silicon rectifier diode whose conductor is controlled by an input current. The schematic symbol for an SCR is shown in Figure 12.21. The symbol is similar to that of a diode with a cathode and an anode. Note that the third element of the SCR is known as the gate. The SCR will conduct current between cathode and anode, but only if the proper control current is applied to the gate. The gate must be made positive with respect to the cathode in order for the SCR to conduct. When conducting, the SCR acts like a closed switch. The voltage drop across the cathode and anode will be approximately 0.7 to 1.8 volts, depending on the size of the SCR and how much current is flowing through it. When the cathode and anode are reverse biased, current will not flow through the device.

Thyristors, like relays, are generally used to control a larger current with a smaller current. Figure 12.22 shows how an SCR is used as a switch to apply DC power to a light bulb. The load could also be a motor or heating element. The DC voltage is connected so that the cathode and anode of the SCR are forward biased, but no current will flow through the device until a current is applied to the gate. This is done with the switch S_1. With the switch open, no current flows in the gate. Closing the switch, however, applies a positive voltage to the gate through resistor R_1. This causes the SCR to turn on. When it conducts, it acts as a low-resistance "on" switch and the light bulb goes on. At this point, switch S_1 may be opened. It is not necessary to maintain current in the gate in order for the device to continue to conduct. The gate current only

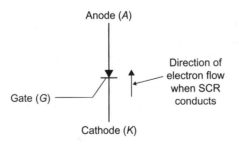

FIGURE 12.21 Schematic symbol of an SCR.

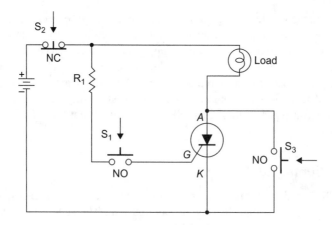

FIGURE 12.22   Circuit that shows how an SCR is turned on and off.

needs to be momentary, for it is only required to turn the device on. The device remains on like a latching relay that has been pulsed.

The gate element is only used to turn the SCR on. Removing gate current will not turn the device off. To cause the SCR to stop conducting, the circuit must be broken. This can be done with a momentary-contact, normally closed pushbutton, in series with the circuit, like S_2 in Figure 12.22. Depressing this pushbutton will break the circuit, stopping the current in the SCR, and the light bulb will go off. To turn the bulb back on, S_1 must again be depressed to apply gate current momentarily.

Another way to stop conduction is to momentarily short around the SCR, as shown in Figure 12.15. When switch S_3 is closed momentarily, current will flow through it and the bulb, bypassing the SCR. The current in the SCR will drop to zero. When switch S_3 is opened, the circuit will be off.

While SCRs are sometimes used to control DC power, in most applications they are used to control AC. Figure 12.23A shows the SCR used to apply AC to a light bulb. If switch S_1 is closed, gate current will be applied to the device. The device, however, will only conduct when the anode is made positive with respect to the cathode. This, of course, occurs when the applied AC voltage has the correct polarity. Since the SCR operates as a rectifier diode, current will only flow through the device on the positive half-cycles of the sine wave. The current through the light bulb will be a pulsating DC, as illustrated in Figure 12.23B. The brightness of the lamp will depend on the average amount of current flowing.

The average amount of current through the light bulb, or other load, can be controlled by using the gate element. Various electronic circuits are used to adjust the time when the SCR turns on. By turning on the SCR at the appropriate time in the positive half-cycle, the duration of the current flow can be controlled. The longer the current is allowed to flow, the higher the average

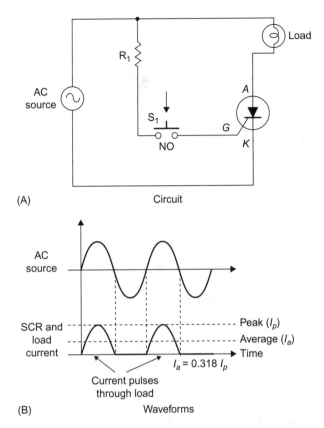

FIGURE 12.23   Using an SCR to control power to a load using AC. The SCR rectifies the AC.

current in the load. By changing the control point, the brightness of the bulb can be varied.

Diac

A diac is a two-terminal semiconductor device that conducts in either direction when a voltage of a specific level is exceeded. This voltage is called the *triggering voltage*. Typical diac triggering voltages are in the 20- to 45-volt range. The diac is used as a triggering device for SCR and triacs as it determines when the device will conduct or cut off.

Triac

A triac is another three-terminal thyristor. Its operation is similar to that of an SCR, but the triac will conduct in either direction. The schematic symbol for the triac is shown in Figure 12.24. Current will flow from main terminal 1 (MT1)

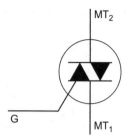

FIGURE 12.24 Schematic symbol for triac.

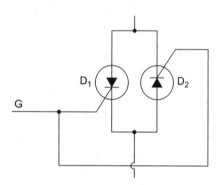

FIGURE 12.25 Equivalent circuit of triac shown with SCRs.

to main terminal 2 (MT2), or from main terminal 2 (MT2) to main terminal 1 (MT1), depending on the polarity of the voltage applied, and if the gate triggers the device on.

An approximate, equivalent circuit of the triac is shown in Figure 12.25. Here, two SCRs are connected back to back with their gates connected together. Current can flow in either direction, depending on whether D_1 or D_2 conducts. Of course, SCR conduction depends on whether a gate-triggering current is applied. Triacs, like SCRs, are used primarily to switch AC voltage to a load. If gate current is applied continuously, the triac will conduct in both forward and backward directions, causing both positive and negative half-cycles of the AC to be applied to a load.

Figure 12.26 shows a circuit used for light-dimming purposes. The load is a light bulb. An RC phase shifter and diac-triggering device are used on the gate of the triac. Assume that the diac has a triggering voltage of 30 volts. If the voltage across the diac is less than 30 volts, it will not conduct. As soon as the 30-volt point is reached or exceeded, the diac rapidly conducts and fires the SCR. Keep in mind that the diac is a bidirectional device, and that current will flow through it in either direction if the bias voltage in either direction exceeds the triggering voltage level.

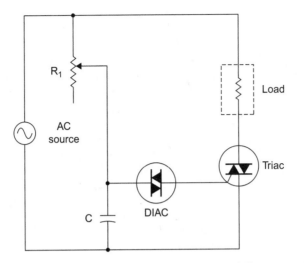

FIGURE 12.26   Lamp dimmer circuit using RC phase shifter and diac to control current in the load.

The triac will conduct on both positive and negative half-cycles of the applied AC. The phase shifter made up of R_1 and C will set the triggering delay. It will determine at what point in the AC cycle the diac will fire and apply gate current to the triac. Figure 12.27 shows the waveforms at two different phase-shift points in the circuit. Note that current flows through the load in both the positive and negative half-cycles, but the duration of current flow depends on the point where the gate triggers. In Figure 12.27, the average current through the load will be relatively high at a 60-degree phase shift, and the bulb will be bright. At 120 degrees, the triggering point is late in each half-cycle, allowing only a small amount of current to flow through the bulb. The bulb, therefore, will glow dimly. Adjusting R_1 will allow the brightness of the bulb to be varied from nearly full on to full off. This is the circuit normally used in household and industrial light dimmers.

This same circuit can be used to control the speed of AC motors. In connecting a hand drill as the load, for example, R_1 will vary the speed of rotation.

Triacs are also widely used as solid-state relays.

PROGRAMMABLE LOGIC CONTROLLERS

A very popular type of controller used in industry and process control is a specialized computer called a programmable logic controller (PLC), or programmable controller. It uses sensors to monitor physical variables and generates output control signals to operate heating elements, motors, pumps, solenoids, and other devices. The PLC, however, contains a microprocessor that can be

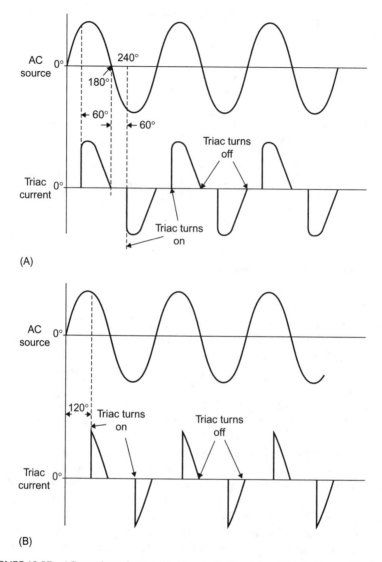

FIGURE 12.27   AC waveforms in a lamp load controlled by a triac. **(A)** Waveforms with triggering at 60 degrees. **(B)** Waveforms with triggering at 120 degrees. The longer pulses produce higher average current in the load.

programmed in a special language that permits the PLC to be customized to the particular control function. In this way, a standard PLC can be programmed and configured for virtually any application. The PLC is by far one of the most popular and versatile control devices in use in industry.

The main function of a PLC is factory automation. It is a versatile special-purpose computer that can be quickly and easily reconfigured to meet changes

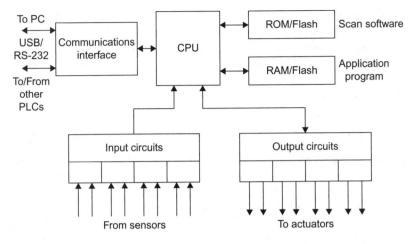

FIGURE 12.28 General block diagram of PLC.

and additions in factories, plants, and other operations that routinely need to control and sequence various machines and processes. While an embedded controller or personal computer could be used in some operations, they are harder to program and apply and are less flexible when rapid changes are needed. Furthermore, these common microcomputers are not designed for the factory environment. But PLCs are rugged and hardened against the harsh environment of most factories and plants. They can withstand the temperature ranges, vibration, and dirty conditions usually existing in a manufacturing or process control setting.

Organization

A general block diagram of a PLC is shown in Figure 12.28. It is like most other microcomputer diagrams but there are some differences. A microprocessor is at the heart of the design. This CPU is coupled to a ROM where a special control program is stored. This program implements the basic scan operational mode of the PLC. RAM is also available to store the application program. Flash memory may also be used in place of RAM.

The key to the PLC's versatility is its extensive I/O section. It has a variety of input and output modules. These are fixed in small, single-function PLCs, but in the larger more flexible units, the PLC is designed to accept a wide variety of separate I/O modules. These usually plug into a rack containing the main PLC unit and power supply. The different I/O modules let the user customize the PLC to the specific application.

Some examples of I/O modules are digital inputs and outputs with common logic levels or conditioned levels, analog-input modules that accept analog signals from sensors, and analog output modules that generate analog outputs for controlling actuators of almost any kind. A huge variety is available.

A communications interface is also available so that the PLC can talk to external devices like a PC or other PLCs that are connected to form a larger control system. RS-232 and RS-485 serial interfaces are common as is USB. The program describing the operations to be implemented is developed on an external PC, and then downloaded to the PLC RAM or flash memory via the USB or RS-232 interfaces.

Operation

The general operation of a PLC is shown in Figure 12.29. When the PLC is started, the internal operational program looks at the application program and interprets what is to be done. The sequence of operations is to first scan the inputs. The PLC looks at what the various sensors are doing. It gets a status update from the application itself. The program then interprets what to do next based on the application program. It processes the input to determine what the outputs should be. Then it updates the outputs to carry out the application. Then the cycle repeats itself at a high rate of speed. The inputs are scanned and outputs updated every 1 millisecond or so, slower or faster as the application requires.

A simple example is using a switch to turn on a motor. The switch contacts are connected to an input module by wires. Output wires connect to a relay that in turn is used to operate the motor. As the PLC is scanning, it sees that the switch is off, so it does nothing. But as soon as the switch is turned

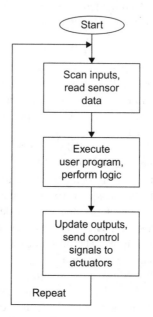

FIGURE 12.29   PLC operational scan cycle.

on, the next scan detects this new condition. The program then tells the CPU that if the switch is on, turn on the relay. An output signal is sent to the output module, which then actuates an external relay. The relay contacts then turn the motor on. If the switch is then turned off, the next scan detects this condition and causes the output to turn off the relay and motor.

Programming

As for programming, PLCs can be programmed in various ways. Most PLC vendors have special languages to do the programming. In some cases, standard BASIC or C languages may be used. One widely used approach is a graphical language called *ladder logic*. Ladder logic is the technology that the PLC replaced decades ago. Control systems were made with relays, and they were wired so the diagram looked like a ladder (see Figure 12.30). The two vertical lines represent a DC or AC voltage source. Circuits are then connected between these two "rails," forming rungs on the ladder. The circuits are connections of switches, relays, and other components. The software allows the user to build the ladder logic program on the screen, while a translation program in the PC converts it into the binary code that the PLC can execute.

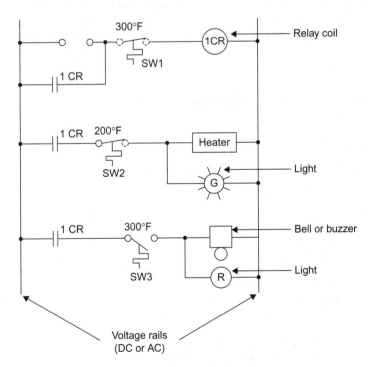

FIGURE 12.30 Sample ladder diagram and PLC application program.

This particular circuit in Figure 12.30 does the following: when the ON pushbutton is pressed, current is applied to relay coil 1CR through normally closed contacts of SW1. SW1 is a thermal switch whose contacts stay closed as long as the temperature is less than 300 degrees Fahrenheit (300°F). If the temperature rises above that level, the contacts open, thereby breaking the circuits. Relay contacts 1CR around the ON pushbutton latch the circuit on when the ON pushbutton is released.

In the next "rung" of the ladder contacts 1CR are on the relay. They close when the relay turns on. This applies voltage to a heating element. This current flows through the normally closed contacts of thermal switch SW2. A green light in parallel with the heating elements turns on to indicate the heater is on.

When the heater temperature reaches 200°F, SW2 opens, turning off the heater and the green light. As the heater cools, SW2 will then close again, applying voltage to the heater. The heater will then cycle off and on keeping its temperature near the 200°F value.

If the temperature should rise above 300°F due to a failure of SW2 °F then SW1 opens. Relay 1CR is turned off, and its contacts open, releasing the latch on rung 1 and opening the circuit on rung 2. Thermal switch SW3 closes at 300°F, turning on a red light and ringing an alarm bell.

This example could be programmed with a graphical ladder logic program or with special commands and syntax of a traditional programming language.

In addition to being able do simple off/off tasks common to industrial control, special hardware in the PLC can also implement counters and timers so that external events can be counted and the sequencing of events can be programmed.

Project 12.1

Build a Robot

Robots are great examples of basic industrial control. While most industrial robots are usually manipulator arms, you can actually buy a mobile robot with wheels that can be programmed for motion and other functions. A wide range of kits are available. It is a great way to get familiar with sensors, a microcomputer controller, and various actuators. Typical inputs are switches, photo cells, temperature sensors, and common outputs are relays, lights, and motors.

Some recommended kits are those made by Lego and Parallax. Suggested websites to learn more follow:

Lego Mindstorms NXT: www.mindstorms.lego.com

Parallax: www.parallax.com

Two magazines regularly cover robots from a hobbyist/experimenter viewpoint. These are *Nuts & Volts* and *Servo*. You may be able to find these at a newsstand; otherwise, go to their websites:

Nuts & Volts Magazine: www.nutsvolts.com

Servo Magazine: www.servomagazine.com

BOOK REFERENCES

Here is a list of books for those who wish to dig deeper. You will find lots of basic electronics books out there, but these are my favorites as well as some of my own.

Ashby D. *Electrical engineering 101*. Newnes/Elsevier; 2008. A good basic book that delves more into the circuits than this book. A good next step from here.

Frenzel LE. *Crash course in electronics technology*. 2nd ed. Newnes/Elsevier; 1996. Out of print now but still available in the used market. A programmed introduction to electronics.

Frenzel LE. *Crash course in digital technology*. 2nd ed. Newnes/Elsevier; 1998. Out of print now but still available in the used market. A programmed introduction to digital circuits. Getting a bit dated now but still valid.

Frenzel LE. *Crash course in PC and microcontroller technology*. Newnes/Elsevier; 1999. Out of print now but still available in the used market. A programmed introduction to micros. Getting dated but the fundamentals are still good.

Frenzel LE. *Principles of electronic communications systems*. 3rd ed. McGraw-Hill; 2008. Recently updated and it takes the systems approach to the subject. A college text but very readable.

Horowitz P, Hill W. *The art of electronics*. 2nd ed. Cambridge University Press; 1989. A bit dated now but a good example of how to present electronics. A great reference. A lab manual is also available. I wish the authors would update this book, but knowing how much work is involved, I might be reluctant too.

Kuphaldt, TR. *Lessons in electric circuits*. Self-published; 2000–2009. A complete electronics book online and free for download. It covers DC and AC circuits, semiconductors and basic circuits, and digital. A lab manual is available. Go to www.openbookproject.new/electricCircuits.

Maxfield C. *Bebop to the Boolean Boogie*. 3rd ed. Newnes/Elsevier; 2009. First published in 1995, it was a hit. A great introduction to digital practice and circuits. It was recently updated.

Michael NA. *Introduction to telecommunication electronics*. 2nd ed. Artech House; 1995. A great little book. Interesting format with each topic presented in two pages. Lots of basic electronic info as well as communications.

doi: 10.1016/B978-1-85617-700-9.00017-5

Index